Lecture Notes in Physics

Edited by H. Araki, Kyoto, J. Ehlers, München, K. Hepp, Zürich
R. Kippenhahn, München, D. Ruelle, Bures-sur-Yvette
H. A. Weidenmüller, Heidelberg, J. Wess, Karlsruhe and J. Zittartz, Köln
Managing Editor: W. Beiglböck

336

Martin Müller

Consistent Classical Supergravity Theories

Springer-Verlag
Berlin Heidelberg GmbH

Author

Martin Müller
Max-Planck-Institut für Physik und Astrophysik
Werner-Heisenberg-Institut für Physik
Föhringer Ring 6, D-8000 München 40, FRG

ISBN 978-3-662-13719-2   ISBN 978-3-540-48114-0 (eBook)
DOI 10.1007/978-3-540-48114-0

Originally published by Springer-Verlag Berlin Heidelberg New York in 1989
Softcover reprint of the hardcover 1st edition 1989

2158/3140-543210 – Printed on acid-free paper

Preface

The purpose of these notes is to give a classification of all consistent supergravity theories in four dimensions, starting from as few assumptions as possible. The only known theories that are not included are the off-shell Poincaré supergravities with central charges or with an infinite number of auxiliary fields. Although most of the results are well known, it should be useful to have a coherent survey without too many loose ends. The reader will find a first orientation in Chapters 1 and 10.

This book is *not* an introduction to supergravity and probably cannot be understood without any previous knowledge of the subject. I would like to recommend the book by Wess and Bagger [1] as an introduction, since I use the same conventions. Some other useful introductory books are listed in Ref. [2]. In principle, however, it should be possible to do all the calculations without consulting any other reference. Parts of these notes may also be used as a starting point for further calculations in extended supergravity. I have tried my best to eliminate all kinds of misprints and I am confident that this goal has been achieved. (However, I apologize in advance for the imperfect English.)

I would like to thank all who contributed directly or indirectly to this book, including P. Binétruy, B. de Wit, N. Dragon, M.-N. Fontaine, D. Franks-Böttcher, G. Girardi, R. Grimm, W. Lang, S. Marculescu, H. Nicolai, C. Ramirez, M. Reuter, J. Rostant, R. Stora, and J. Wess. In particular, I thank Julius Wess for the opportunity to give several lectures on supergravity at the University of Karlsruhe. I am also grateful to U. Baur and W. Lerche for their help with TeX and to Springer-Verlag for a TeX macro package. Most of this work was done while I was a Fellow of the Theory Division of CERN.

Munich, April 1989   Martin Müller

Contents

Part I: Conformal Supergravity

1. Introduction and Summary 1
   1.1 Brief Survey 1
   1.2 Assumptions 2
   1.3 Results 2

2. Superspace Geometry 4
   2.1 Affine Structure 4
   2.2 Metric Structure 5
   2.3 Explicit Equations 6
   2.4 Transformation Laws 6

3. Constraints 8

4. Bianchi Identities 10

5. Symmetries of the Constraints 20

6. Consistency of the Constraints 23

7. Linearized Ricci Identities 25

8. Non-Linear Ricci Identities 34

9. Invariant Actions 37
   9.1 Chiral Superfields 37
   9.2 Superfield Actions 38
   9.3 Chiral Actions 39
   9.4 Summary 45

Part II: Poincaré Supergravity

10. Introduction and Summary 47
    10.1 Brief Survey 47
    10.2 Assumptions 48
    10.3 Results 48

11. Upper Bounds 50

12. On-Shell Poincaré Supergravity 51
    12.1 Linearized $N = 8$ Supergravity 51
    12.2 Non-Linear $N = 8$ Supergravity 53

13. One-Form Gauge Potentials 56
    13.1 Superspace Geometry 56
    13.2 Constraints 57
    13.3 Bianchi Identities 58
    13.4 Ricci Identities 60
    13.5 $N = 4$ 63
    13.6 $N = 3, 4$ Off-Shell Supergravity 66

14. Two-Form Gauge Potentials 67
    14.1 Superspace Geometry 67
    14.2 Constraints 68
    14.3 Bianchi Identities 68
    14.4 Ricci Identities 70
    14.5 Solution of the Constraints 71

15. $N = 2$ Off-Shell Supergravity 73
    15.1 The Minimal Field Representation 73
    15.2 Multiplets Without Local SO(2) 75
    15.3 Multiplets with Local SO(2) 78

16. $N = 1$ Off-Shell Supergravity 84
    16.1 The Minimal Multiplet 84
    16.2 The New Minimal Multiplet 85

Part III: Conclusion

17. Matter Couplings 87
    17.1 $N = 1$ 88
    17.2 $N = 2$ 93
    17.3 $N > 2$ 97

18. Outlook into Quantum Supergravity 98
    18.1 Poincaré Supergravity 98
    18.2 Conformal Supergravity 98

Appendix A: Conventions 99
Appendix B: Differential Forms 101
Appendix C: Useful Formulas 102
Appendix D: Non-Linear Ricci Identities 108
Appendix E: Second Bianchi Identity 114
Appendix F: Non-Linear $N = 8$ Supergravity 117

References 121
Super-Index 125

Part I

Conformal Supergravity

1. Introduction and Summary

1.1 Brief Survey

According to the theorem by Haag, Łopuszański, and Sohnius [3], the most general graded Lie algebra of a massive quantum field theory is the Poincaré supersymmetry algebra. Its generators are:

$$\begin{array}{ll} P_m & \text{translations} \\ M_{[mn]} & \text{Lorentz transformations} \\ Q_\mu^M & \text{supersymmetry transformations} \\ Z^{[MN]} & \text{central charges} \\ B_i & \text{compact Lie group } \mathcal{G} \end{array}$$

($m = 0, \ldots, 3$ is a space-time index, $\mu = 1, 2$ is a Weyl spinor index, $M = 1, \ldots, N$ and $i = 1, \ldots, n$ are internal indices.)

A massless theory may have even more symmetries: Poincaré supersymmetry with dilatations or conformal supersymmetry. The superconformal algebra has no central charges, but it contains in addition the following generators:

$$\begin{array}{ll} D & \text{dilatations} \\ K_m & \text{special conformal transformations} \\ S_M^\mu & \text{special supersymmetry transformations} \end{array}$$

The internal symmetry group is $\mathcal{G} = \mathcal{G}' \times \mathcal{G}''$, where $\mathcal{G}'$ acts on the capital indices and $\mathcal{G}''$ leaves them invariant. Moreover, one has $\mathcal{G}' = \mathrm{U}(N)$ for $N \neq 4$ and $\mathcal{G}' = \mathrm{U}(4)$ or $\mathrm{SU}(4)$ for $N = 4$.

The gauge theory of supersymmetry is supergravity. Since both translations and special conformal transformations get absorbed into general coordinate transformations, there are only two possibilities: Poincaré supergravity and conformal supergravity.

In the first part of these notes we shall be concerned with conformal supergravities, which are generalizations of Weyl's theory of gravity. They are very attractive from a theoretical point of view because they have more symmetries than any other field theory. In particular, the conformal symmetry constrains all masses to vanish and all fundamental constants to be dimensionless. With this in mind, however, a look out of the window into

1

nature is rather frustrating. Not only are there lots of massive particles in the world, but there is also a long-range force, namely gravitation, with a dimensional coupling constant. It seems, therefore, that the only role of conformal supergravity can be to clarify the structure of Poincaré supergravity. Nonetheless it is not completely excluded that the conformal symmetry plays a more important part at very high energies (see also Chapter 18).

Conformal supergravity theories were found from $N = 1$ up to $N = 4$ (see Ref. [4] for a review). All of them can be extended off the mass-shell. The $N = 1$ theory was constructed as the gauge theory of the superconformal group [5], the $N = 2$ multiplet was discovered as a subset of an off-shell Poincaré multiplet [6], and the $N = 4$ theory was derived from the multiplet of currents [7]. For $N = 3$ and $N = 4$, however, only the linearized actions are known. In Ref. [8] it was argued that off-shell theories do not exist for $N > 4$ because they would contain fields with spin > 2. This was confirmed in Ref. [9] by an explicit derivation of conformal field equations for $N > 4$. On the other hand, $N > 4$ on-shell conformal supergravity is not ruled out by these arguments. Thus the situation for $N > 4$ is rather unclear.

In the following we shall try to rederive the above results in a more systematic way and to close the gaps that are still left. To this aim we shall employ the superspace approach invented by Wess and Zumino [10,1] and further developed by Howe [11,12]. Since this is our main assumption (see below), we have tried to make the book self-consistent up to some basic conventions. It may therefore also serve as a review on the differential geometry of extended superspace.

1.2 Assumptions

Our classification of consistent conformal supergravity theories is based on the following two assumptions:

(A) The theory can be formulated in conventional extended superspace (Wess-Zumino superspace).

(B) Two covariant (super)fields with dimension zero and high spins vanish (Eq. (3.9)).

These assumptions are satisfied by all known theories and it is very likely (although impossible to prove) that they cannot be circumvented.

1.3 Results

Under the above assumptions, our results can be summarized in the following table:

N	1	2	3	4	5	≥ 6
off-shell	8	24	64	128	—	—
on-shell	8	20	48	96	256	—

Table 1. Consistent conformal supergravity theories

The numbers are the bosonic (fermionic) degrees of freedom of the minimal multiplets.

2

By "consistent" we mean (i) for *off-shell* theories: the supergravity algebra closes on the fields and there exists an invariant action, and (ii) for *on-shell* theories: the algebra closes on the fields and the field equations. Note that we do not require the existence of "on-shell actions". For $N \leq 4$ these are simply the off-shell actions without auxiliary fields, whereas for $N = 5$ an invariant action does not exist.

With the exception of $N = 4$, all the theories in Table 1 are unique. The $N = 4$ multiplet and its transformation laws are unique, too, but the most general action includes an arbitrary holomorphic function.

Furthermore, the $N \neq 4$ theories have a local $U(N)$ symmetry, while for $N = 4$ there are only gauge fields corresponding to $SU(4)$. The $N = 4$ transformation laws are in addition invariant under a local $U(1)$ and a global $SU(1,1)$ [7], but the action breaks the $SU(1,1)$ invariance.

2. Superspace Geometry

2.1 Affine Structure

The coordinates of conventional extended superspace are

$$z^{\mathcal{M}} = \left(x^m, \ \theta^{\mu}_M, \ \bar{\theta}^{M}_{\dot{\mu}} \right), \tag{2.1}$$

where $M = 1, \ldots, N$ is an internal index. The biggest and most convenient structure group is $\mathrm{SL}(2, \mathbf{C}) \times \mathrm{U}(N)$ with the parameters

$$L_{\mathcal{B}}{}^{\mathcal{A}} = \left(L_b{}^a, \ L^{B\alpha}_{\beta A}, \ L^{\dot{\beta} A}_{B\dot{\alpha}} \right), \tag{2.2}$$

where

$$
\begin{aligned}
L^{B\alpha}_{\beta A} &= \delta^B_A \, L_\beta{}^\alpha + \delta^\alpha_\beta \, L^B{}_A \,, \\
L^{\dot{\beta} A}_{B\dot{\alpha}} &= \delta^A_B \, L^{\dot{\beta}}{}_{\dot{\alpha}} - \delta^{\dot{\beta}}_{\dot{\alpha}} \, L^A{}_B \,,
\end{aligned} \tag{2.3}
$$

and

$$
\begin{aligned}
L_{ba} &= -L_{ab}\,, \quad L_{\beta\alpha} = L_{\alpha\beta}\,, \quad L_{\dot{\beta}\dot{\alpha}} = L_{\dot{\alpha}\dot{\beta}}\,, \\
L_{\beta\dot{\beta}\,\alpha\dot{\alpha}} &= 2\,\varepsilon_{\dot{\beta}\dot{\alpha}}\,L_{\beta\alpha} - 2\,\varepsilon_{\beta\alpha}\,L_{\dot{\beta}\dot{\alpha}}\,.
\end{aligned} \tag{2.4}
$$

The $\mathrm{U}(N)$ can be further decomposed into $\mathrm{SU}(N) \times \mathrm{U}(1)$:

$$L^B{}_A = \widetilde{L}^B{}_A + \delta^B_A \, L\,, \qquad \widetilde{L}^A{}_A = 0\,. \tag{2.5}$$

Another manifest symmetry of superspace is the general coordinate transformations, which include both general coordinate transformations of space-time and local supersymmetry transformations. The remaining two symmetries of conformal supergravity, dilatations and special supersymmetry transformations, will appear in Chapter 5 within the so-called super-Weyl transformations.

An arbitrary p-form $\Omega^{\mathcal{A}}$ and its exterior derivative transform under the structure group as follows:

$$\delta\Omega^{\mathcal{A}} = \Omega^{\mathcal{B}} \, L_{\mathcal{B}}{}^{\mathcal{A}}\,, \tag{2.6}$$

$$\delta \mathrm{d}\Omega^{\mathcal{A}} = \left(\mathrm{d}\Omega^{\mathcal{B}} \right) L_{\mathcal{B}}{}^{\mathcal{A}} + \Omega^{\mathcal{B}} \, \mathrm{d}L_{\mathcal{B}}{}^{\mathcal{A}}\,. \tag{2.7}$$

In order to get a homogeneous transformation law, one defines the covariant exterior derivative

$$\mathcal{D}\Omega^{\mathcal{A}} = \mathrm{d}\Omega^{\mathcal{A}} + \Omega^{\mathcal{B}} \, \Phi_{\mathcal{B}}{}^{\mathcal{A}}\,. \tag{2.8}$$

The connection

$$\Phi_{\mathcal{B}}{}^{\mathcal{A}} = \mathrm{d}z^{\mathcal{M}} \, \Phi_{\mathcal{M}\mathcal{B}}{}^{\mathcal{A}} \tag{2.9}$$

is a Lie algebra valued 1-form, i.e., it has the properties (2.2–5).

4

Its transformation law

$$\delta \Phi_B{}^A = -\mathcal{D} L_B{}^A \tag{2.10}$$

implies the desired result

$$\delta \, \mathcal{D}\Omega^A = \left(\mathcal{D}\Omega^B\right) L_B{}^A. \tag{2.11}$$

The curvature is defined by

$$\mathcal{D}\mathcal{D}\Omega^A = \Omega^B R_B{}^A, \tag{2.12}$$

$$R_B{}^A = \mathrm{d}\Phi_B{}^A + \Phi_B{}^C \Phi_C{}^A. \tag{2.13}$$

It is a Lie algebra valued 2-form,

$$R_B{}^A = \tfrac{1}{2} \, \mathrm{d}z^M \, \mathrm{d}z^N R_{NMB}{}^A, \tag{2.14}$$

and satisfies the Bianchi identity

$$\mathcal{D} R_B{}^A = 0. \tag{2.15}$$

2.2 Metric Structure

In the next step, one chooses an "orthonormal" basis of 1-forms, the vielbein

$$E^A = \mathrm{d}z^M E_M{}^A. \tag{2.16}$$

Its determinant

$$E = \det E_M{}^A \neq 0 \tag{2.17}$$

is defined via det = exp tr log, where tr is the supertrace. The inverse vielbein is defined by

$$E_A{}^M E_M{}^B = \delta_A^B,$$
$$E_M{}^A E_A{}^N = \delta_M^N. \tag{2.18}$$

The covariant derivative of the vielbein,

$$T^A = \mathcal{D} E^A, \tag{2.19}$$

$$T^A = \tfrac{1}{2} \, \mathrm{d}z^M \, \mathrm{d}z^N T_{NM}{}^A = \tfrac{1}{2} E^B E^C T_{CB}{}^A, \tag{2.20}$$

is the torsion 2-form, which satisfies the Bianchi identity

$$\mathcal{D} T^A = E^B R_B{}^A. \tag{2.21}$$

2.3 Explicit Equations

Of course, all the equations for differential forms can also be written in terms of component fields, either in the basis $dz^{\mathcal{M}}$ of coordinate differentials or in the vielbein basis $E^{\mathcal{A}}$. For instance, the covariant derivative (2.8) of an arbitrary 0-form $V_{\mathcal{A}}$ is

$$\mathcal{D}_{\mathcal{M}} V_{\mathcal{A}} = \partial_{\mathcal{M}} V_{\mathcal{A}} - \Phi_{\mathcal{M}\mathcal{A}}{}^{\mathcal{B}} V_{\mathcal{B}} \tag{2.22}$$

and the (anti)commutator of two covariant derivatives (2.12) gives

$$[\mathcal{D}_{\mathcal{A}}, \mathcal{D}_{\mathcal{B}}\} V_{\mathcal{C}} = -T_{\mathcal{A}\mathcal{B}}{}^{\mathcal{D}} \mathcal{D}_{\mathcal{D}} V_{\mathcal{C}} - R_{\mathcal{A}\mathcal{B}\mathcal{C}}{}^{\mathcal{D}} V_{\mathcal{D}} . \tag{2.23}$$

Furthermore, we shall need the structure equations (2.19, 13)

$$T_{\mathcal{N}\mathcal{M}}{}^{\mathcal{A}} = \sum_{\mathcal{N}\mathcal{M}} \mathcal{D}_{\mathcal{N}} E_{\mathcal{M}}{}^{\mathcal{A}}, \tag{2.24}$$

$$R_{\mathcal{N}\mathcal{M}\mathcal{B}}{}^{\mathcal{A}} = \sum_{\mathcal{N}\mathcal{M}} (\partial_{\mathcal{N}} \Phi_{\mathcal{M}\mathcal{B}}{}^{\mathcal{A}} - \Phi_{\mathcal{N}\mathcal{B}}{}^{\mathcal{C}} \Phi_{\mathcal{M}\mathcal{C}}{}^{\mathcal{A}}) \tag{2.25}$$

and the Bianchi identities (2.21, 15)

$$\sum_{\mathcal{D}\mathcal{C}\mathcal{B}} (R_{\mathcal{D}\mathcal{C}\mathcal{B}}{}^{\mathcal{A}} - \mathcal{D}_{\mathcal{D}} T_{\mathcal{C}\mathcal{B}}{}^{\mathcal{A}} - T_{\mathcal{D}\mathcal{C}}{}^{\mathcal{E}} T_{\mathcal{E}\mathcal{B}}{}^{\mathcal{A}}) = 0, \tag{2.26}$$

$$\sum_{\mathcal{E}\mathcal{D}\mathcal{C}} (\mathcal{D}_{\mathcal{E}} R_{\mathcal{D}\mathcal{C}\mathcal{B}}{}^{\mathcal{A}} + T_{\mathcal{E}\mathcal{D}}{}^{\mathcal{F}} R_{\mathcal{F}\mathcal{C}\mathcal{B}}{}^{\mathcal{A}}) = 0 \tag{2.27}$$

in their explicit form. The operator $\sum$ and other conventions are explained in Appendix A.

2.4 Transformation Laws

Finally, we consider the transformation properties of the various component fields under the structure group and under general coordinate transformations with parameters $\delta z^{\mathcal{M}} = \xi^{\mathcal{M}}$. For an arbitrary 0-form $V^{\mathcal{A}}$, the vielbein $E_{\mathcal{M}}{}^{\mathcal{A}}$, and the connection $\Phi_{\mathcal{M}\mathcal{B}}{}^{\mathcal{A}}$, one finds

$$\delta V^{\mathcal{A}} = \xi^{\mathcal{M}} \partial_{\mathcal{M}} V^{\mathcal{A}} + V^{\mathcal{B}} L_{\mathcal{B}}{}^{\mathcal{A}}, \tag{2.28}$$

$$\delta E_{\mathcal{M}}{}^{\mathcal{A}} = \xi^{\mathcal{N}} \partial_{\mathcal{N}} E_{\mathcal{M}}{}^{\mathcal{A}} + (\partial_{\mathcal{M}} \xi^{\mathcal{N}}) E_{\mathcal{N}}{}^{\mathcal{A}} + E_{\mathcal{M}}{}^{\mathcal{B}} L_{\mathcal{B}}{}^{\mathcal{A}}, \tag{2.29}$$

$$\delta \Phi_{\mathcal{M}\mathcal{B}}{}^{\mathcal{A}} = \xi^{\mathcal{N}} \partial_{\mathcal{N}} \Phi_{\mathcal{M}\mathcal{B}}{}^{\mathcal{A}} + (\partial_{\mathcal{M}} \xi^{\mathcal{N}}) \Phi_{\mathcal{N}\mathcal{B}}{}^{\mathcal{A}} - \mathcal{D}_{\mathcal{M}} L_{\mathcal{B}}{}^{\mathcal{A}}. \tag{2.30}$$

With the new parameters

$$\xi^{\mathcal{A}} = \xi^{\mathcal{M}} E_{\mathcal{M}}{}^{\mathcal{A}},$$

$$\Lambda_{\mathcal{B}}{}^{\mathcal{A}} = L_{\mathcal{B}}{}^{\mathcal{A}} - \xi^{\mathcal{M}} \Phi_{\mathcal{M}\mathcal{B}}{}^{\mathcal{A}} \tag{2.31}$$

the above transformation laws can be written in the more covariant form

$$\delta V^A = \xi^B \mathcal{D}_B V^A + V^B \Lambda_B{}^A, \tag{2.32}$$

$$\delta E_\mathcal{M}{}^A = \xi^B T_{B\mathcal{M}}{}^A + \mathcal{D}_\mathcal{M} \xi^A + E_\mathcal{M}{}^B \Lambda_B{}^A, \tag{2.33}$$

$$\delta \Phi_{\mathcal{M}B}{}^A = \xi^C R_{C\mathcal{M}B}{}^A - \mathcal{D}_\mathcal{M} \Lambda_B{}^A. \tag{2.34}$$

The supergravity algebra, i.e., the commutator of two such transformations, reads

$$\left[\delta(\hat{\xi}^A, \hat{\Lambda}_B{}^A), \delta(\xi^A, \Lambda_B{}^A) \right] =$$

$$= \delta\left(-\xi^B \hat{\xi}^C T_{CB}{}^A, \xi^C \hat{\xi}^D R_{DCB}{}^A + [\Lambda, \hat{\Lambda}]_B{}^A \right). \tag{2.35}$$

3. Constraints

Consider arbitrary variations of the vielbein and the connection:

$$\delta E^{A} = H^{A}, \tag{3.1}$$

$$\delta \Phi_{B}{}^{A} = \Omega_{B}{}^{A}. \tag{3.2}$$

These variations change the torsion and the curvature as follows:

$$\delta T^{A} = \mathcal{D}H^{A} + E^{B}\Omega_{B}{}^{A}, \tag{3.3}$$

$$\delta R_{B}{}^{A} = \mathcal{D}\Omega_{B}{}^{A}. \tag{3.4}$$

In the vielbein basis, these equations read explicitly

$$\delta T_{CB}{}^{A} = \sum_{CB}\left(\mathcal{D}_{C}H_{B}{}^{A} + \Omega_{CB}{}^{A} - H_{C}{}^{D}T_{DB}{}^{A}\right) + T_{CB}{}^{D}H_{D}{}^{A}, \tag{3.5}$$

$$\delta R_{DCB}{}^{A} = \sum_{DC}\left(\mathcal{D}_{D}\Omega_{CB}{}^{A} - H_{D}{}^{\mathcal{E}}R_{\mathcal{E}CB}{}^{A}\right) + T_{DC}{}^{\mathcal{E}}\Omega_{\mathcal{E}B}{}^{A}. \tag{3.6}$$

The various components of the fields occuring in the above equations are listed in the following table:

dim	vielbein	connection	torsion	curvature
$-\frac{1}{2}$	$H_{\underline{\beta}}{}^{a}$			
0	$H_{b}{}^{a}, H_{\underline{\beta}}{}^{\underline{\alpha}}$		$T_{\underline{\gamma}\,\underline{\beta}}{}^{a}$	
$\frac{1}{2}$	$H_{b}{}^{\underline{\alpha}}$	$\Omega_{\underline{\gamma}B}{}^{A}$	$T_{\underline{\gamma}b}{}^{a}, T_{\underline{\gamma}\,\underline{\beta}}{}^{\underline{\alpha}}$	
1		$\Omega_{cB}{}^{A}$	$T_{cb}{}^{a}, T_{c\underline{\beta}}{}^{\underline{\alpha}}$	$R_{\underline{\delta}\,\underline{\gamma}B}{}^{A}$
$\frac{3}{2}$			$T_{cb}{}^{\underline{\alpha}}$	$R_{\underline{\delta}cB}{}^{A}$
2				$R_{dcB}{}^{A}$

Table 2. Components of $H_{B}{}^{A}$, $\Omega_{CB}{}^{A}$, torsion, and curvature

It is easy to see that at dimensions 0, $\frac{1}{2}$, and 1 certain components of the torsion and the curvature can be absorbed by redefinitions of the vielbein and the connection. We shall discuss this in more detail in the following.

dim 0

At dimension 0 we have the torsion components $T_{\gamma\,\beta}^{CB\,a}$, $T_{\gamma B}^{C\dot\beta\,a}$, and their complex conjugates. First we split off the value of $T_{\gamma B}^{C\dot\beta\,a}$ in flat superspace:

$$T_{\gamma B}^{C\dot\beta\,a} = 2i\,\delta_B^C\,(\sigma^a)_\gamma^{\ \dot\beta} + \widehat{T}_{\gamma B}^{C\dot\beta\,a}. \tag{3.7}$$

Then we exploit our freedom to redefine the vielbein arbitrarily. Using (3.5), one finds that the only parts of the torsion which cannot be absorbed by $H_b^{\ a}$ and $H_\beta^{\ \alpha}$ are [13]

$$T_{\gamma\,\beta\,\alpha\dot\alpha}^{CB} = T_{(\gamma\beta\alpha)\dot\alpha}^{(CB)},$$

$$\widehat{T}_{\gamma\,\dot\beta B\,\alpha\dot\alpha}^{C} = \widetilde{T}_{(\gamma\alpha)(\dot\beta\dot\alpha)B}^{C}, \tag{3.8}$$

where $\widetilde{T}$ is traceless in CB. Since it would not be possible to describe any dynamics in the presence of these covariant high-spin fields, we set them equal to zero:

$$T_{(\gamma\beta\alpha)\dot\alpha}^{(CB)} = 0,$$

$$\widetilde{T}_{(\gamma\alpha)(\dot\beta\dot\alpha)B}^{C} = 0. \tag{3.9}$$

It should be emphasized that these are the only equations which have to be postulated in the superspace approach. All the other constraints are purely "conventional" ones, i. e., they correspond to redefinitions of the vielbein and the connection.

dim $\frac{1}{2}$

At dimension $\frac{1}{2}$ we can use $H_b^{\ \alpha}$ and $\Omega_{\underline{\gamma}B}^{\ \ A}$ to absorb parts of $T_{\underline{\gamma}b}^{\ \ a}$ and $T_{\underline{\gamma}\underline{\beta}}^{\ \ \alpha}$. This leads, for example, to

$$T_{\gamma\,\beta\dot\beta\,\alpha\dot\alpha}^{C} = T_{(\gamma\beta\alpha)(\dot\beta\dot\alpha)}^{C},$$

$$T_{\gamma B\dot\alpha}^{C\dot\alpha A} = 0. \tag{3.10}$$

dim 1

At dimension 1, a suitable redefinition of the Lorentz connection gives

$$T_{cb}^{\ \ a} = 0, \tag{3.11}$$

which is the well-known constraint of Riemannian geometry. Moreover, we can use $\Omega_c^{\ B}_{\ A}$ to absorb parts of $T_{c\underline{\beta}}^{\ \ \alpha}$ and $R_{\delta\,\underline{\gamma}B}^{\ \ \ A}$. At this point, however, the most convenient choice of this constraint is not obvious, so we defer it to the next chapter.

The above constraints eliminate most of the components of vielbein and connection as independent variables. These restrictions and their consequences will be investigated in Chapters 5 and 6. In particular, we shall discuss the consistency of the constraints (3.9), which is not at all clear. In the next chapter we shall first analyze the consequences for the covariant tensors torsion and curvature.

4. Bianchi Identities

The components of the torsion and the curvature are not independent fields, but they are related by the Bianchi identities (2.26–27). Therefore the constraints on the lowest-dimensional components affect the higher-dimensional ones, too. To analyze this in a systematic way, we start with the first Bianchi identity (2.26), which consists of 30 equations from dimension $\frac{1}{2}$ up to dimension $\frac{5}{2}$. In the following we shall denote these identities by their indices $\left(\mathcal{D}_{CB}{}^{A} \right)$ and we shall almost always omit complex conjugate equations.

dim $\frac{1}{2}$: $\left({}_{\delta\gamma\beta}{}^{a} \right)$

Using the dim-0 constraints

$$T_{\gamma\beta}^{CBa} = 0,$$

$$T_{\gamma B}^{C\dot\beta a} = 2i\,\delta_B^C\,(\sigma^a)_\gamma{}^{\dot\beta}, \tag{4.1}$$

one obtains from $\left({}^{DCBa}_{\delta\gamma\beta} \right)$

$$T_{\gamma\beta\dot\alpha}^{CBA} = \varepsilon_{\gamma\beta}\,\overline{W}_{\dot\alpha}^{[CBA]}, \tag{4.2}$$

whereby $\overline{W}$ is defined. The second dim-$\frac{1}{2}$ identity $\left({}^{DC\dot\beta a}_{\delta\gamma B} \right)$ together with the constraints (3.10) yields

$$T_{\gamma b}^{C\ a} = T_{\gamma\beta A}^{CB\alpha} = T_{\gamma B\dot\alpha}^{C\dot\beta A} = 0. \tag{4.3}$$

PROOF: Let us show as an example how the first result (4.2) can be derived. The relevant Bianchi identity reads

$$\sum_{\mathcal{D}CB} \left(R_{\delta\gamma\beta}^{DCBa} - \mathcal{D}_\delta^D\,T_{\gamma\beta}^{CBa} - T_{\delta\gamma}^{DC\mathcal{E}}\,T_{\mathcal{E}\beta}^{Ba} \right) = 0.$$

Using (2.2) and the constraints (4.1), one obtains

$$(\sigma^a)_\beta{}^{\dot\varepsilon}\,T_{\delta\gamma\dot\varepsilon}^{DCB} + (\sigma^a)_\delta{}^{\dot\varepsilon}\,T_{\gamma\beta\dot\varepsilon}^{CBD} + (\sigma^a)_\gamma{}^{\dot\varepsilon}\,T_{\beta\delta\dot\varepsilon}^{BDC} = 0.$$

To get rid of the vector index, we multiply this equation by $(\sigma_a)_{\alpha\dot\alpha}$:

$$\varepsilon_{\alpha\beta}\,T_{\delta\gamma\dot\alpha}^{DCB} + \varepsilon_{\alpha\delta}\,T_{\gamma\beta\dot\alpha}^{CBD} + \varepsilon_{\alpha\gamma}\,T_{\beta\delta\dot\alpha}^{BDC} = 0. \tag{a}$$

Here we used (C.29). Next we contract with $\varepsilon^{\beta\alpha}$ and find

$$2\,T_{\delta\gamma\dot\alpha}^{DCB} + T_{\gamma\delta\dot\alpha}^{CBD} + T_{\gamma\delta\dot\alpha}^{BDC} = 0. \tag{b}$$

Now we decompose $T_{\gamma\beta\dot\alpha}^{CBA}$ into

$$T_{\gamma\beta\dot\alpha}^{CBA} = T_{(\gamma\beta)\dot\alpha}^{(CB)A} + \varepsilon_{\gamma\beta}\,T_{\dot\alpha}^{[CB]A}.$$

Equation (b) then splits into

$$2T^{(DC)B}_{(\delta\gamma)\dot\alpha} \dot+ T^{(CB)D}_{(\delta\gamma)\dot\alpha} + T^{(BD)C}_{(\delta\gamma)\dot\alpha} = 0,$$ (c)

$$2T^{[DC]B}_{\dot\alpha} - T^{[CB]D}_{\dot\alpha} - T^{[BD]C}_{\dot\alpha} = 0.$$ (d)

The part of (c) which is completely symmetric in DCB reads

$$T^{(DC)B}_{(\delta\gamma)\dot\alpha} + T^{(CB)D}_{(\delta\gamma)\dot\alpha} + T^{(BD)C}_{(\delta\gamma)\dot\alpha} = 0.$$

Subtracting this from (c) yields

$$T^{(DC)B}_{(\delta\gamma)\dot\alpha} = 0.$$

Equation (d) can be written as

$$T^{[DC]B}_{\dot\alpha} = \tfrac{1}{3}\left(T^{[DC]B}_{\dot\alpha} + T^{[CB]D}_{\dot\alpha} + T^{[BD]C}_{\dot\alpha}\right)$$

$$\equiv \overline{W}^{[DCB]}_{\dot\alpha}.$$

Altogether, we obtain

$$T^{CBA}_{\gamma\beta\dot\alpha} = \varepsilon_{\gamma\beta}\,\overline{W}^{[CBA]}_{\dot\alpha}.$$

The original identity (a) is then satisfied because of (C.4). This completes the proof of (4.2).

dim 1: $\left(\underline{{}_{\delta\gamma b}}{}^{a}\right),\ \left(\underline{{}_{\delta\gamma\beta}}{}^{\alpha}\right)$

At dimension 1 we obtain from $\left({}^{DC}_{\delta\gamma b}{}^{a}\right)$

$$T^{BA}_{\gamma\dot\gamma\,\beta\dot\alpha} = \mathrm{i}\left(\varepsilon_{\dot\gamma\dot\alpha}\,V^{[BA]}_{(\gamma\beta)} + \varepsilon_{\gamma\beta}\,\overline{W}^{[BA]}_{(\dot\gamma\dot\alpha)} + \varepsilon_{\gamma\beta}\,\varepsilon_{\dot\gamma\dot\alpha}\,S^{(BA)}\right),$$ (4.4)

$$R^{DC}_{\delta\gamma\beta\alpha} = -2\left(\varepsilon_{\delta\gamma}\,V^{DC}_{\beta\alpha} + \sum_{\beta\alpha}\varepsilon_{\delta\beta}\,\varepsilon_{\gamma\alpha}\,S^{DC}\right),$$ (4.5)

$$R^{DC}_{\delta\gamma\beta\dot\alpha} = 2\varepsilon_{\delta\gamma}\,\overline{W}^{DC}_{\beta\dot\alpha}$$ (4.6)

and from $\left({}^{DC\,\dot\beta A}_{\delta\gamma\,B\dot\alpha}\right)$

$$\mathcal{D}_{\dot\beta D}\,\overline{W}^{CBA}_{\dot\alpha} - \sum_{CBA}\delta^{C}_{D}\,\overline{W}^{BA}_{\dot\beta\dot\alpha} + \varepsilon_{\dot\beta\dot\alpha}\,\overline{M}^{[CBA]}_{D},$$ (4.7)

$$R^{DCB}_{\delta\gamma\,A} = 2\sum_{DC}^{\widehat{}}\delta^{D}_{A}\,V^{CB}_{\delta\gamma} + 2\varepsilon_{\delta\gamma}\sum_{DC}\delta^{D}_{A}\,S^{CB} - \varepsilon_{\delta\gamma}\,\overline{M}^{DCB}_{A}.$$ (4.8)

$\left(\begin{smallmatrix} D C B \alpha \\ \delta \gamma \beta A \end{smallmatrix}\right)$ is then identically satisfied. Next we consider $\left(\begin{smallmatrix} D \dot{\gamma} & a \\ \delta C b & \end{smallmatrix}\right)$,

$$T_{\gamma\dot{\gamma}\,\beta\,\alpha A}{}^{B} = \frac{i}{2}\left(\varepsilon_{\beta\alpha}\,T_{\gamma\dot{\gamma}\,A}^{B} + \sum_{\beta\alpha}\varepsilon_{\gamma\beta}\,U_{\alpha\dot{\gamma}\,A}^{B}\right),$$

$$\dot{T}_{\gamma\dot{\gamma}\,\beta B\,\dot{\alpha}}{}^{A} = \frac{i}{2}\left(\varepsilon_{\dot{\beta}\dot{\alpha}}\,T_{\gamma\dot{\gamma}\,B}^{A} + \sum_{\dot{\beta}\dot{\alpha}}\varepsilon_{\dot{\gamma}\dot{\beta}}\,U_{\gamma\dot{\alpha}\,B}^{A}\right), \tag{4.9}$$

$$R_{\delta\dot{\gamma}C\,\beta\alpha}^{D} = \sum_{\beta\alpha}\varepsilon_{\delta\beta}\,U_{\alpha\dot{\gamma}\,C}^{D},$$

$$R_{\delta\dot{\gamma}C\,\dot{\beta}\dot{\alpha}}^{D} = \sum_{\dot{\beta}\dot{\alpha}}\varepsilon_{\dot{\gamma}\dot{\beta}}\,U_{\delta\dot{\alpha}\,C}^{D}, \tag{4.10}$$

and $\left(\begin{smallmatrix} D \dot{\gamma} B \alpha \\ \delta C \beta A \end{smallmatrix}\right)$,

$$R_{\delta\dot{\gamma}C}{}^{B}{}_{A} = \delta_{C}^{D}\,T_{\delta\dot{\gamma}\,A}^{B} - \left(\delta_{C}^{D}\,U_{\delta\dot{\gamma}\,A}^{B} + \delta_{A}^{B}\,U_{\delta\dot{\gamma}\,C}^{D}\right)$$
$$+ 2\left(\delta_{A}^{D}\,U_{\delta\dot{\gamma}\,C}^{B} + \delta_{C}^{B}\,U_{\delta\dot{\gamma}\,A}^{D}\right) + W_{\delta CAE}\,\overline{W}_{\dot{\gamma}}^{DBE}. \tag{4.11}$$

Now we are in a position to choose the conventional constraint corresponding to $\Omega_{c}{}^{B}{}_{A}$ in (3.5–6), i. e., to a redefinition of the U(N) connection. The most convenient choice is

$$T_{\alpha\dot{\alpha}\,A}^{B} = U_{\alpha\dot{\alpha}\,A}^{B}, \tag{4.12}$$

which simplifies (4.9) and (4.11) considerably:

$$T_{\gamma\dot{\gamma}\,\beta\,\alpha A}{}^{B} = i\,\varepsilon_{\gamma\alpha}\,U_{\beta\dot{\gamma}\,A}^{B},$$

$$T_{\gamma\dot{\gamma}\,\beta B\,\dot{\alpha}}{}^{A} = i\,\varepsilon_{\dot{\gamma}\dot{\alpha}}\,U_{\gamma\dot{\beta}\,B}^{A}, \tag{4.13}$$

$$R_{\delta\dot{\gamma}C}{}^{B}{}_{A} = 2\left(\delta_{A}^{D}\,U_{\delta\dot{\gamma}\,C}^{B} + \delta_{C}^{B}\,U_{\delta\dot{\gamma}\,A}^{D}\right) - \delta_{A}^{B}\,U_{\delta\dot{\gamma}\,C}^{D}$$
$$+ W_{\delta CAE}\,\overline{W}_{\dot{\gamma}}^{DBE}. \tag{4.14}$$

The last dim-1 identity, $\left(\begin{smallmatrix} D C B A \\ \delta \gamma \beta \dot{\alpha} \end{smallmatrix}\right)$, gives

$$\mathcal{D}_{\alpha}^{D}\,\overline{W}_{\dot{\alpha}}^{CBA} = i\,\overline{N}_{\alpha\dot{\alpha}}^{[DCBA]}. \tag{4.15}$$

Let us summarize the results obtained so far. All the components of torsion and curvature can be expressed in terms of the basic superfields

$$\overline{W}_{\dot{\alpha}}^{CBA},$$

$$V_{\beta\alpha}^{BA}, \quad S^{BA}, \quad U_{\alpha\dot{\alpha}\,A}^{B},$$

their complex conjugates, and their covariant derivatives. (We shall see that this holds for the higher-dimensional components, too.) For $N = 2$ and $N = 1$, $\overline{W}_{\dot{\alpha}}^{CBA}$ is replaced by $\overline{W}_{\dot{\beta}\dot{\alpha}}^{BA}$ and $\overline{W}_{\dot{\gamma}\dot{\beta}\dot{\alpha}}^{A}$, respectively.

12

dim $\frac{3}{2}$: $\left(\underline{\ }_{\delta c b}{}^{a}\right),\ \left(\underline{\ }_{\delta c\,\beta}{}^{\alpha}\right)$

At dimension $\frac{3}{2}$ we start by decomposing the torsion component $T_{cb\,\dot\alpha}{}^{A}$:

$$T_{\gamma\dot\gamma\,\beta\dot\beta\,\dot\alpha}{}^{A} = \varepsilon_{\dot\gamma\dot\beta}\,\overline{\Psi}_{(\gamma\beta)\dot\alpha}^{A} + \varepsilon_{\gamma\beta}\Big(\overline{W}_{(\dot\gamma\dot\beta\dot\alpha)}^{A} + \sum_{\dot\gamma\dot\beta}\varepsilon_{\dot\gamma\dot\alpha}\,\overline{\Psi}_{\dot\beta}^{A}\Big). \tag{4.16}$$

$\left(\underline{\ }_{\delta c b}^{D}{}^{a}\right)$ is then equivalent to

$$R_{\delta\,\gamma\dot\gamma\,\beta\alpha}^{D} = \mathrm{i}\,\varepsilon_{\delta\gamma}\,\overline{\Psi}_{\beta\alpha\,\dot\gamma}^{D} + \frac{\mathrm{i}}{2}\sum_{\beta\alpha}\varepsilon_{\delta\beta}\big(\overline{\Psi}_{\gamma\alpha\,\dot\gamma}^{D} - 3\,\varepsilon_{\gamma\alpha}\,\overline{\Psi}_{\dot\gamma}^{D}\big), \tag{4.17}$$

$$R_{\delta\,\gamma\dot\gamma\,\dot\beta\dot\alpha}^{D} = -2\mathrm{i}\,\varepsilon_{\delta\gamma}\,\overline{W}_{\dot\gamma\dot\beta\dot\alpha}^{D} + \frac{\mathrm{i}}{2}\sum_{\dot\beta\dot\alpha}\varepsilon_{\dot\gamma\dot\beta}\big(\overline{\Psi}_{\delta\gamma\,\dot\alpha}^{D} + \varepsilon_{\delta\gamma}\,\overline{\Psi}_{\dot\alpha}^{D}\big) \tag{4.18}$$

and $\left(\underline{\ }_{\delta c B\dot\alpha}^{D\,\dot\beta A}\right)$ implies

$$\mathcal{D}_{\dot\gamma C}\,\overline{W}_{\dot\beta\dot\alpha}^{BA} = -2\sum_{BA}\delta_{C}^{B}\,\overline{W}_{\dot\gamma\dot\beta\dot\alpha}^{A} - \frac{1}{3!}\sum_{\dot\gamma\dot\beta\dot\alpha}\overline{V}_{\dot\gamma\dot\beta\,CD}\,\overline{W}_{\dot\alpha}^{DBA} + \sum_{\dot\beta\dot\alpha}\varepsilon_{\dot\gamma\dot\beta}\,\overline{\Lambda}_{\dot\alpha C}^{[BA]}, \tag{4.19}$$

$$\begin{aligned}\mathcal{D}_{\beta}^{C}\,U_{\alpha\dot\alpha\,A}^{B} = &-\mathcal{D}_{\dot\alpha A}\,V_{\beta\alpha}^{CB} - 2\delta_{A}^{C}\,\overline{\Psi}_{\beta\alpha\,\dot\alpha}^{B} + \delta_{A}^{B}\,\overline{\Psi}_{\beta\alpha\,\dot\alpha}^{C} - W_{\beta\alpha\,AD}\,\overline{W}_{\dot\alpha}^{DCB}\\ &+ \varepsilon_{\beta\alpha}\big(\mathcal{D}_{\dot\alpha A}\,S^{CB} - \overline{\Lambda}_{\dot\alpha A}^{CB} + 2\delta_{A}^{C}\,\overline{\Psi}_{\dot\alpha}^{B} + \delta_{A}^{B}\,\overline{\Psi}_{\dot\alpha}^{C}\\ &- \frac{1}{3}\overline{V}_{\dot\alpha\dot\beta\,AD}\,\overline{W}^{\dot\beta DCB} + \overline{S}_{AD}\,\overline{W}_{\dot\alpha}^{DCB}\big),\end{aligned} \tag{4.20}$$

$$\begin{aligned}R_{\delta\,\gamma\dot\gamma\,A}^{D\quad B} = &2\mathrm{i}\,\delta_{A}^{D}\,\overline{\Psi}_{\delta\gamma\,\dot\gamma}^{B} - \frac{\mathrm{i}}{2}\delta_{A}^{B}\,\overline{\Psi}_{\delta\gamma\,\dot\gamma}^{D} + \mathrm{i}\,W_{\delta\gamma\,AE}\,\overline{W}_{\dot\gamma}^{EDB}\\ &+ \mathrm{i}\,\varepsilon_{\delta\gamma}\big(2\,\overline{\Lambda}_{\dot\gamma A}^{DB} + 2\delta_{A}^{D}\,\overline{\Psi}_{\dot\gamma}^{B} - \frac{1}{2}\delta_{A}^{B}\,\overline{\Psi}_{\dot\gamma}^{D}\\ &- \frac{1}{3}\overline{V}_{\dot\gamma\dot\epsilon\,AE}\,\overline{W}^{\dot\epsilon EDB} - \overline{S}_{AE}\,\overline{W}_{\dot\gamma}^{EDB}\big).\end{aligned} \tag{4.21}$$

Thereupon $\left(\underline{\ }_{\delta c\,\beta A}^{D\ B\alpha}\right)$ is identically satisfied and $\left(\underline{\ }_{\delta c\,\beta\,\dot\alpha}^{D\ BA}\right)$ gives

$$\mathcal{D}_{\alpha}^{C}\,\overline{W}_{\dot\beta\dot\alpha}^{BA} = -\frac{1}{2}\sum_{\dot\beta\dot\alpha}\big(\mathrm{i}\,\mathcal{D}_{\alpha\dot\beta}\,\overline{W}_{\dot\alpha}^{CBA} + U_{\alpha\dot\beta\,D}^{C}\,\overline{W}_{\dot\alpha}^{DBA}\big), \tag{4.22}$$

$$\begin{aligned}\mathcal{D}_{\gamma}^{C}\,V_{\beta\alpha}^{BA} = &\,V_{(\gamma\beta\alpha)}^{[CBA]} + \frac{1}{6}\sum_{\beta\alpha}\sum_{BA}\varepsilon_{\gamma\beta}\big(2\,\mathcal{D}_{\alpha}^{B}\,S^{CA} + \frac{\mathrm{i}}{2}\,\mathcal{D}_{\alpha\dot\alpha}\,\overline{W}^{\dot\alpha CBA}\\ &- \frac{1}{2}\,U_{\alpha\dot\alpha\,D}^{C}\,\overline{W}^{\dot\alpha DBA} - 2\,U_{\alpha\dot\alpha\,D}^{B}\,\overline{W}^{\dot\alpha DCA}\big),\end{aligned} \tag{4.23}$$

$$\widehat{\sum_{CBA}}\,\mathcal{D}_{\alpha}^{C}S^{BA} = 0. \tag{4.24}$$

dim 2: $\left(_{dcb}{}^a\right)$, $\left(_{dc\underline{\beta}}{}^\alpha\right)$

The first dim-2 identity, $\left(_{dcb}{}^a\right)$, is well known from Riemannian geometry. It reads

$$\sum_{dcb} R_{dcb}{}^a = 0,\tag{4.25}$$

which is equivalent to

$$R_{dc\,ba} = R_{ba\,dc}\,,\qquad \varepsilon^{abcd}\, R_{dc\,ba} = 0\,.\tag{4.26}$$

In spinor notation, this becomes

$$R_{\delta\dot{\delta}\,\gamma\dot{\gamma}\,\beta\alpha} = \varepsilon_{\delta\gamma}\, P_{(\beta\alpha)(\dot{\delta}\dot{\gamma})} + \varepsilon_{\dot{\delta}\dot{\gamma}}\Big(W_{(\delta\gamma\beta\alpha)} + \frac{1}{12}\sum_{\beta\alpha}\varepsilon_{\delta\beta}\,\varepsilon_{\gamma\alpha}\,R\Big),$$

$$R_{\delta\dot{\delta}\,\gamma\dot{\gamma}\,\dot{\beta}\dot{\alpha}} = -\varepsilon_{\dot{\delta}\dot{\gamma}}\, P_{(\delta\gamma)(\dot{\beta}\dot{\alpha})} - \varepsilon_{\delta\gamma}\Big(\overline{W}_{(\dot{\delta}\dot{\gamma}\dot{\beta}\dot{\alpha})} + \frac{1}{12}\sum_{\dot{\beta}\dot{\alpha}}\varepsilon_{\dot{\delta}\dot{\beta}}\,\varepsilon_{\dot{\gamma}\dot{\alpha}}\,R\Big),\tag{4.27}$$

where $R = R_{ab}{}^{ab}$. Similarly, the U(N) curvature $R_{dc}{}^B{}_A$ can be decomposed into

$$R_{\delta\dot{\delta}\,\gamma\dot{\gamma}}{}^B{}_A = \varepsilon_{\dot{\delta}\dot{\gamma}}\,\rho_{(\delta\gamma)A}^{\ B} - \varepsilon_{\delta\gamma}\,\overline{\rho}_{(\dot{\delta}\dot{\gamma})A}^{\ B}\,.\tag{4.28}$$

The remaining two identities, $\left(_{dc B\dot{\alpha}}{}^{\dot{\beta}A}\right)$ and $\left(_{dc\,\beta}{}^{BA}{}_{\dot{\alpha}}\right)$, determine the covariant spinor derivatives of $T_{cb}{}^A_{\ \dot{\alpha}}$:

$$\mathcal{D}_B^{\dot{\beta}}\, T_{dc}{}^A_{\ \dot{\alpha}} = R_{dc\,B\dot{\alpha}}^{\ \ \dot{\beta}A} - \sum_{dc}\big(\mathcal{D}_d\, T_{cB\dot{\alpha}}^{\ \ \dot{\beta}A} + T_{dB}^{\ \ \dot{\beta}\underline{\varepsilon}}\, T_{c\underline{\varepsilon}}{}^A_{\ \dot{\alpha}}\big),\tag{4.29}$$

$$\mathcal{D}_\beta^B\, T_{dc}{}^A_{\ \dot{\alpha}} = -\sum_{dc}\big(\mathcal{D}_d\, T_{c\,\beta\ \dot{\alpha}}^{\ BA} + T_{d\beta}^{\ B\underline{\varepsilon}}\, T_{c\underline{\varepsilon}}{}^A_{\ \dot{\alpha}}\big) - T_{dcE}^{\ \ \ \varepsilon}\, T^{EBA}_{\ \ \varepsilon\,\beta\,\dot{\alpha}}\,.\tag{4.30}$$

Equation (4.29) implies

$$\mathcal{D}_{\dot{\beta}B}\,\overline{\Psi}^A_{\beta\alpha\dot{\alpha}} = \frac{1}{2}\sum_{\beta\alpha}\Big[-\delta^A_B\, P_{\beta\alpha\,\dot{\beta}\dot{\alpha}} + \mathrm{i}\,\mathcal{D}_{\alpha\dot{\alpha}}\, U^A_{\beta\dot{\beta}\,B} - U^A_{\alpha\dot{\alpha}\,C}\, U^C_{\beta\dot{\beta}\,B}$$

$$+ \overline{V}_{\dot{\beta}\dot{\alpha}\,BC}\, V^{CA}_{\beta\alpha} + W_{\beta\alpha\,BC}\,\overline{W}^{CA}_{\dot{\beta}\dot{\alpha}} + \varepsilon_{\dot{\beta}\dot{\alpha}}\big(-\rho^A_{\beta\alpha\,B}$$

$$+ W_\beta{}^\gamma{}_{BC}\, V^{CA}_{\gamma\alpha} + W_{\beta\alpha\,BC}\, S^{CA} - \overline{S}_{BC}\, V^{CA}_{\beta\alpha}\big)\Big],\tag{4.31}$$

$$\mathcal{D}_{\dot{\delta}B}\,\overline{W}^A_{\dot{\gamma}\dot{\beta}\dot{\alpha}} = -\delta^A_B\,\overline{W}_{\dot{\delta}\dot{\gamma}\dot{\beta}\dot{\alpha}} + \frac{1}{3!}\sum_{\dot{\gamma}\dot{\beta}\dot{\alpha}}\Big[2\overline{V}_{\dot{\delta}\dot{\gamma}\,BC}\,\overline{W}^{CA}_{\dot{\beta}\dot{\alpha}}$$

$$+ \varepsilon_{\dot{\delta}\dot{\gamma}}\big(\overline{\rho}^A_{\dot{\beta}\dot{\alpha}\,B} - 2\overline{S}_{BC}\,\overline{W}^{CA}_{\dot{\beta}\dot{\alpha}}\big)\Big],\tag{4.32}$$

$$\mathcal{D}_{\dot{\beta}B}\,\overline{\Psi}^A_{\dot{\alpha}} = \frac{1}{3}\overline{\rho}^A_{\dot{\beta}\dot{\alpha}\,B} - \frac{\mathrm{i}}{2}\,\mathcal{D}^\alpha_{\ \dot{\alpha}}\, U^A_{\alpha\dot{\beta}\,B} - \frac{1}{2}\, U^{\alpha A}_{\ \ \dot{\alpha}\,C}\, U^C_{\alpha\dot{\beta}\,B}$$

$$- \frac{1}{3}\overline{V}_{\dot{\beta}\dot{\gamma}\,BC}\,\overline{W}^{\dot{\gamma}\,CA}_{\ \ \dot{\alpha}} + \overline{V}_{\dot{\beta}\dot{\alpha}\,BC}\, S^{CA} + \frac{1}{3}\overline{S}_{BC}\,\overline{W}^{CA}_{\dot{\beta}\dot{\alpha}}$$

$$+ \frac{1}{2}\varepsilon_{\dot{\beta}\dot{\alpha}}\Big(\frac{1}{6}\delta^A_B\, R - W^{\alpha\beta}_{BC}\, V^{CA}_{\alpha\beta} - 2\overline{S}_{BC}\, S^{CA}\Big)\tag{4.33}$$

and (4.30) gives

$$\mathcal{D}^B_\gamma \overline{\Psi}^A_{\beta\alpha\dot\alpha} = \frac{1}{2} \sum_{\beta\alpha} \Big[\mathrm{i}\, \mathcal{D}_{\alpha\dot\alpha}\, V^{BA}_{\gamma\beta} + U^A_{\alpha\dot\alpha\, C}\, V^{CB}_{\gamma\beta} + U^B_{\gamma\dot\alpha\, C}\, V^{CA}_{\beta\alpha}$$

$$+ W_{\gamma\beta\alpha\, C}\, \overline{W}^{CBA}_{\dot\alpha} + \varepsilon_{\gamma\beta}\, \big(\mathrm{i}\, \mathcal{D}_\alpha^{\ \dot\beta}\, \overline{W}^{BA}_{\dot\beta\dot\alpha} - \mathrm{i}\, \mathcal{D}_{\alpha\dot\alpha}\, S^{BA}$$

$$- U^{A\dot\beta}_{\alpha\, C}\, \overline{W}^{CB}_{\dot\beta\dot\alpha} + U^A_{\alpha\dot\alpha\, C}\, S^{CB} - 2\, \Psi_{\alpha C}\, \overline{W}^{CBA}_{\dot\alpha} \big) \Big], \tag{4.34}$$

$$\mathcal{D}^B_\alpha \overline{W}^A_{\dot\gamma\dot\beta\dot\alpha} = \frac{1}{3!} \sum_{\dot\gamma\dot\beta\dot\alpha} \big(\mathrm{i}\, \mathcal{D}_{\alpha\dot\gamma}\, \overline{W}^{BA}_{\dot\beta\dot\alpha} + 2\, U^B_{\alpha\dot\gamma\, C}\, \overline{W}^{CA}_{\dot\beta\dot\alpha} + \Psi_{\dot\gamma\dot\beta\,\alpha C}\, \overline{W}^{CBA}_{\dot\alpha} \big), \tag{4.35}$$

$$\mathcal{D}^B_\alpha \overline{\Psi}^A_{\dot\alpha} = -\frac{\mathrm{i}}{2}\, \mathcal{D}^\beta_{\ \dot\alpha}\, V^{BA}_{\beta\alpha} + \frac{\mathrm{i}}{6}\, \mathcal{D}_\alpha^{\ \dot\beta}\, \overline{W}^{BA}_{\dot\beta\dot\alpha} + \frac{\mathrm{i}}{2}\, \mathcal{D}_{\alpha\dot\alpha}\, S^{BA}$$

$$+ \frac{1}{2}\, U^{\beta A}_{\dot\alpha\, C}\, V^{CB}_{\beta\alpha} + \frac{1}{3}\, U^{B\dot\beta}_{\alpha\, C}\, \overline{W}^{CA}_{\dot\beta\dot\alpha} - \frac{1}{2}\, U^{A\dot\beta}_{\alpha\, C}\, \overline{W}^{CB}_{\dot\beta\dot\alpha}$$

$$+ U^B_{\alpha\dot\alpha\, C}\, S^{CA} + \frac{1}{2}\, U^A_{\alpha\dot\alpha\, C}\, S^{CB} - \frac{1}{3}\, \Psi_{\dot\alpha\dot\beta\,\alpha C}\, \overline{W}^{\dot\beta CBA}. \tag{4.36}$$

dim $\frac{5}{2}$: $\left({}_{dcb}{}^\alpha \right)$

At dimension $\frac{5}{2}$ we are left with the identity

$$\sum_{dcb} \big(\mathcal{D}_d\, T_{cb}^{\ \ A}{}_{\dot\alpha} - T_{dc}^{\ \ \varepsilon}\, T_{b\underline{\varepsilon}}^{\ \ A}{}_{\dot\alpha} \big) = 0, \tag{4.37}$$

which is equivalent to

$$\sum_{\dot\beta\dot\alpha} \big(\mathcal{D}^\beta_{\ \dot\beta}\, \overline{\Psi}^A_{\beta\alpha\dot\alpha} - \mathcal{D}_\alpha^{\ \dot\gamma}\, \overline{W}^A_{\dot\gamma\dot\beta\dot\alpha} + \mathcal{D}_{\alpha\dot\beta}\, \overline{\Psi}^A_{\dot\alpha} + \mathrm{i}\, U^{A\dot\gamma}_{\alpha B}\, \overline{W}^B_{\dot\gamma\dot\beta\dot\alpha}$$

$$+ 2\mathrm{i}\, U^A_{\alpha\dot\beta B}\, \overline{\Psi}^B_{\dot\alpha} + \mathrm{i}\, \Psi_{\dot\beta\dot\alpha B}^{\ \ \ \beta}\, V^{BA}_{\beta\alpha} + \mathrm{i}\, \Psi_{\dot\beta\dot\gamma\,\alpha B}\, \overline{W}^{\dot\gamma BA}_{\dot\alpha}$$

$$+ \mathrm{i}\, \Psi_{\dot\beta\dot\alpha\,\alpha B}\, S^{BA} + 3\mathrm{i}\, \Psi_{\alpha B}\, \overline{W}^{BA}_{\dot\beta\dot\alpha} \big) = 0, \tag{4.38}$$

$$\mathcal{D}^{\beta\dot\alpha}\, \overline{\Psi}^A_{\beta\alpha\dot\alpha} - 3\, \mathcal{D}_\alpha^{\ \dot\alpha}\, \overline{\Psi}^A_{\dot\alpha} - 2\mathrm{i}\, U^{\beta\dot\alpha A}_{\ \ \ B}\, \overline{\Psi}^B_{\beta\alpha\dot\alpha} - \mathrm{i}\, \Psi_{\dot\alpha\dot\beta\,\alpha B}\, \overline{W}^{\dot\alpha\dot\beta BA}$$

$$+ 2\mathrm{i}\, W_{\alpha\beta\gamma B}\, V^{\beta\gamma BA} - 2\mathrm{i}\, \Psi^\beta_B\, V^{BA}_{\beta\alpha} + 6\mathrm{i}\, \Psi_{\alpha B}\, S^{BA} = 0. \tag{4.39}$$

This completes the solution of the first Bianchi identity. The second Bianchi identity (2.27) causes much less problems. In fact, the following theorem shows that it does not contain any new information.

THEOREM (Dragon [14]): The second Bianchi identity follows from the first Bianchi identity and the Ricci identity (2.12).

PROOF: Applying $\mathcal{D}$ to the first Bianchi identity (2.21) and using the Ricci identity (2.12) gives $E^B\, \mathcal{D}R_B{}^A = 0$. With the definition $I_B{}^A = \mathcal{D}R_B{}^A$, this can be written in

15

component form as

$$\sum_{\mathcal{EDCB}} I_{\mathcal{EDCB}}{}^{\mathcal{A}} = 0.$$

Since $I_B{}^{\mathcal{A}}$ is Lie algebra valued, this implies (after a simple calculation) $I_{\mathcal{EDCB}}{}^{\mathcal{A}} = 0$ or $I_B{}^{\mathcal{A}} = 0$, which is the second Bianchi identity (2.15).

Summary

Torsion:

$$T_{\gamma B}^{C\dot\beta a} = 2\mathrm{i}\,\delta_B^C\,(\sigma^a)_\gamma{}^{\dot\beta} \tag{4.40}$$

$$T_{\gamma\beta\dot\alpha}^{CBA} = \varepsilon_{\gamma\beta}\,\overline{W}_{\dot\alpha}^{CBA}$$

$$T_{\dot\gamma C\dot\beta B\alpha A} = -\varepsilon_{\dot\gamma\dot\beta}\,W_{\alpha CBA} \tag{4.41}$$

$$T_{\gamma\dot\gamma\,\beta\dot\alpha}^{BA} = \mathrm{i}\left(\varepsilon_{\dot\gamma\dot\alpha}\,V_{\gamma\beta}^{BA} + \varepsilon_{\gamma\beta}\,\overline{W}_{\dot\gamma\dot\alpha}^{BA} + \varepsilon_{\gamma\beta}\,\varepsilon_{\dot\gamma\dot\alpha}\,S^{BA}\right)$$

$$T_{\gamma\dot\gamma\,\dot\beta B\alpha A} = \mathrm{i}\left(\varepsilon_{\gamma\alpha}\,\overline{V}_{\dot\gamma\dot\beta BA} + \varepsilon_{\dot\gamma\dot\beta}\,W_{\gamma\alpha BA} + \varepsilon_{\gamma\alpha}\,\varepsilon_{\dot\gamma\dot\beta}\,\overline{S}_{BA}\right) \tag{4.42}$$

$$T_{\gamma\dot\gamma\,\beta\alpha A}^{B} = \mathrm{i}\,\varepsilon_{\gamma\alpha}\,U_{\beta\dot\gamma A}^{B}$$

$$T_{\gamma\dot\gamma\,\dot\beta B\dot\alpha}^{A} = \mathrm{i}\,\varepsilon_{\dot\gamma\dot\alpha}\,U_{\gamma\dot\beta B}^{A} \tag{4.43}$$

$$T_{\gamma\dot\gamma\,\beta\dot\beta\dot\alpha}^{A} = \varepsilon_{\dot\gamma\dot\beta}\,\overline{\Psi}_{\gamma\beta\dot\alpha}^{A} + \varepsilon_{\gamma\beta}\left(\overline{W}_{\dot\gamma\dot\beta\dot\alpha}^{A} + \sum_{\dot\gamma\dot\beta}\varepsilon_{\dot\gamma\dot\alpha}\,\overline{\Psi}_{\dot\beta}^{A}\right)$$

$$T_{\gamma\dot\gamma\,\beta\dot\beta\alpha A} = \varepsilon_{\gamma\beta}\,\Psi_{\dot\gamma\dot\beta\alpha A} + \varepsilon_{\dot\gamma\dot\beta}\left(W_{\gamma\beta\alpha A} + \sum_{\gamma\beta}\varepsilon_{\gamma\alpha}\,\Psi_{\beta A}\right) \tag{4.44}$$

Lorentz curvature:

$$R_{\delta\gamma\beta\alpha}^{DC} = -2\left(\varepsilon_{\delta\gamma}\,V_{\beta\alpha}^{DC} + \sum_{\beta\alpha}\varepsilon_{\delta\beta}\,\varepsilon_{\gamma\alpha}\,S^{DC}\right)$$

$$R_{\dot\delta D\dot\gamma C\dot\beta\dot\alpha} = -2\left(\varepsilon_{\dot\delta\dot\gamma}\,\overline{V}_{\dot\beta\dot\alpha DC} + \sum_{\dot\beta\dot\alpha}\varepsilon_{\dot\delta\dot\beta}\,\varepsilon_{\dot\gamma\dot\alpha}\,\overline{S}_{DC}\right) \tag{4.45}$$

$$R_{\delta\gamma\dot\beta\dot\alpha}^{DC} = 2\varepsilon_{\delta\gamma}\,\overline{W}_{\dot\beta\dot\alpha}^{DC}$$

$$R_{\dot\delta D\dot\gamma C\beta\alpha} = 2\varepsilon_{\dot\delta\dot\gamma}\,W_{\beta\alpha DC} \tag{4.46}$$

$$R_{\delta\dot\gamma C\beta\alpha}^{D} = \sum_{\beta\alpha}\varepsilon_{\delta\beta}\,U_{\alpha\dot\gamma C}^{D}$$

$$R_{\delta\dot\gamma C\dot\beta\dot\alpha}^{D} = \sum_{\dot\beta\dot\alpha}\varepsilon_{\dot\gamma\dot\beta}\,U_{\delta\dot\alpha C}^{D} \tag{4.47}$$

$$R^{D}_{\delta\gamma\dot\gamma\beta\alpha} = i\,\varepsilon_{\delta\gamma}\,\overline{\Psi}^{D}_{\beta\alpha\dot\gamma} + \frac{i}{2}\sum_{\beta\alpha}\varepsilon_{\delta\beta}\left(\overline{\Psi}^{D}_{\gamma\alpha\dot\gamma} - 3\varepsilon_{\gamma\alpha}\overline{\Psi}^{D}_{\dot\gamma}\right)$$

$$R_{\dot\delta D\gamma\dot\gamma\dot\beta\dot\alpha} = i\,\varepsilon_{\dot\delta\dot\gamma}\,\Psi_{\dot\beta\dot\alpha\gamma D} + \frac{i}{2}\sum_{\dot\beta\dot\alpha}\varepsilon_{\dot\delta\dot\beta}\left(\Psi_{\dot\gamma\dot\alpha\gamma D} - 3\varepsilon_{\dot\gamma\dot\alpha}\Psi_{\gamma D}\right) \tag{4.48}$$

$$R^{D}_{\delta\gamma\dot\gamma\dot\beta\dot\alpha} = -2i\,\varepsilon_{\delta\gamma}\,\overline{W}^{D}_{\dot\gamma\dot\beta\dot\alpha} + \frac{i}{2}\sum_{\dot\beta\dot\alpha}\varepsilon_{\dot\gamma\dot\beta}\left(\overline{\Psi}^{D}_{\delta\gamma\dot\alpha} + \varepsilon_{\delta\gamma}\overline{\Psi}^{D}_{\dot\alpha}\right)$$

$$R_{\dot\delta D\gamma\dot\gamma\beta\alpha} = -2i\,\varepsilon_{\dot\delta\dot\gamma}\,W_{\gamma\beta\alpha D} + \frac{i}{2}\sum_{\beta\alpha}\varepsilon_{\gamma\beta}\left(\Psi_{\dot\delta\dot\gamma\alpha D} + \varepsilon_{\dot\delta\dot\gamma}\Psi_{\alpha D}\right) \tag{4.49}$$

$$R_{\delta\dot\delta\,\gamma\dot\gamma\beta\alpha} = \varepsilon_{\delta\gamma}\,P_{\beta\alpha\dot\delta\dot\gamma} + \varepsilon_{\dot\delta\dot\gamma}\left(W_{\delta\gamma\beta\alpha} + \frac{1}{12}\sum_{\beta\alpha}\varepsilon_{\delta\beta}\varepsilon_{\gamma\alpha}R\right)$$

$$R_{\delta\dot\delta\,\gamma\dot\gamma\dot\beta\dot\alpha} = -\varepsilon_{\dot\delta\dot\gamma}\,P_{\delta\gamma\dot\beta\dot\alpha} - \varepsilon_{\delta\gamma}\left(\overline{W}_{\dot\delta\dot\gamma\dot\beta\dot\alpha} + \frac{1}{12}\sum_{\dot\beta\dot\alpha}\varepsilon_{\dot\delta\dot\beta}\varepsilon_{\dot\gamma\dot\alpha}R\right) \tag{4.50}$$

U(N) curvature:

$$R^{DCB}_{\delta\gamma\;A} = 2\sum_{DC}\widehat{\delta}^{D}_{A}V^{CB}_{\delta\gamma} + 2\varepsilon_{\delta\gamma}\sum_{DC}\delta^{D}_{A}S^{CB} - \varepsilon_{\delta\gamma}\overline{M}^{DCB}_{A}$$

$$R_{\dot\delta D\dot\gamma C\;A}^{\quad\quad\;\;B} = 2\sum_{DC}\widehat{\delta}^{B}_{D}\overline{V}_{\dot\delta\dot\gamma CA} + 2\varepsilon_{\dot\delta\dot\gamma}\sum_{DC}\delta^{B}_{D}\overline{S}_{CA} - \varepsilon_{\dot\delta\dot\gamma}M^{B}_{DCA} \tag{4.51}$$

$$R^{D}_{\delta\dot\gamma C\;A}{}^{B} = 2\left(\delta^{D}_{A}U^{B}_{\delta\dot\gamma C} + \delta^{B}_{C}U^{D}_{\delta\dot\gamma A}\right) - \delta^{B}_{A}U^{D}_{\delta\dot\gamma C}$$
$$\quad\quad + W_{\delta CAE}\overline{W}^{DBE}_{\dot\gamma} \tag{4.52}$$

$$R^{D}_{\delta\gamma\dot\gamma\;A}{}^{B} = 2i\,\delta^{D}_{A}\overline{\Psi}^{B}_{\delta\gamma\dot\gamma} - \frac{i}{2}\delta^{B}_{A}\overline{\Psi}^{D}_{\delta\gamma\dot\gamma} + i\,W_{\delta\gamma AE}\overline{W}^{EDB}_{\dot\gamma}$$
$$\quad\quad + i\,\varepsilon_{\delta\gamma}\left(2\overline{\Lambda}^{DB}_{\dot\gamma A} + 2\delta^{D}_{A}\overline{\Psi}^{B}_{\dot\gamma} - \frac{1}{2}\delta^{B}_{A}\overline{\Psi}^{D}_{\dot\gamma}\right.$$
$$\quad\quad \left. - \frac{1}{3}\overline{V}_{\dot\gamma\dot\varepsilon AE}\overline{W}^{\dot\varepsilon EDB} - \overline{S}_{AE}\overline{W}^{EDB}_{\dot\gamma}\right)$$

$$R_{\dot\delta D\gamma\dot\gamma\;A}^{\quad\quad\;\;B} = 2i\,\delta^{B}_{D}\Psi_{\dot\delta\dot\gamma\gamma A} - \frac{i}{2}\delta^{B}_{A}\Psi_{\dot\delta\dot\gamma\gamma D} + i\,\overline{W}^{BE}_{\dot\delta\dot\gamma}W_{\gamma EDA}$$
$$\quad\quad + i\,\varepsilon_{\dot\delta\dot\gamma}\left(2\Lambda^{B}_{\gamma DA} + 2\delta^{B}_{D}\Psi_{\gamma A} - \frac{1}{2}\delta^{B}_{A}\Psi_{\gamma D}\right.$$
$$\quad\quad \left. - \frac{1}{3}V^{BE}_{\gamma\varepsilon}W^{\varepsilon}_{EDA} - S^{BE}W_{\gamma EDA}\right) \tag{4.53}$$

$$R_{\delta\dot\delta\,\gamma\dot\gamma\;A}^{\quad\quad\;B} = \varepsilon_{\dot\delta\dot\gamma}\,\rho^{B}_{\delta\gamma A} - \varepsilon_{\delta\gamma}\,\overline{\rho}^{B}_{\dot\delta\dot\gamma A} \tag{4.54}$$

17

Thus all the components of torsion and curvature can be expressed in terms of a few superfields, which are related by covariant conditions. Up to dimension $\frac{3}{2}$, these conditions are: [1]

$$\mathcal{D}_{\dot{\beta}D}\,\overline{W}_{\dot{\alpha}}^{CBA} = \sum_{CBA} \delta_D^C\,\overline{W}_{\dot{\beta}\dot{\alpha}}^{BA} + \varepsilon_{\dot{\beta}\dot{\alpha}}\,\overline{M}_D^{CBA}$$

$$\mathcal{D}_\beta^D\,W_{\alpha\,CBA} = -\sum_{CBA} \delta_C^D\,W_{\beta\alpha\,BA} - \varepsilon_{\beta\alpha}\,M_{CBA}^D \qquad (4.55)$$

$$\mathcal{D}_\alpha^D\,\overline{W}_{\dot{\alpha}}^{CBA} = \mathrm{i}\,\overline{N}_{\alpha\dot{\alpha}}^{DCBA}$$

$$\mathcal{D}_{\dot{\alpha}D}\,W_{\alpha\,CBA} = \mathrm{i}\,N_{\alpha\dot{\alpha}\,DCBA} \qquad (4.56)$$

$$\mathcal{D}_{\dot{\gamma}C}\,\overline{W}_{\dot{\beta}\dot{\alpha}}^{BA} = -2\sum_{BA} \delta_C^B\,\overline{W}_{\dot{\gamma}\dot{\beta}\dot{\alpha}}^A - \frac{1}{3!}\sum_{\dot{\gamma}\dot{\beta}\dot{\alpha}} \overline{V}_{\dot{\gamma}\dot{\beta}\,CD}\,\overline{W}_{\dot{\alpha}}^{DBA}$$
$$+ \sum_{\dot{\beta}\dot{\alpha}} \varepsilon_{\dot{\gamma}\dot{\beta}}\,\overline{\Lambda}_{\dot{\alpha}C}^{BA}$$

$$\mathcal{D}_\gamma^C\,W_{\beta\alpha\,BA} = -2\sum_{BA} \delta_B^C\,W_{\gamma\beta\alpha\,A} - \frac{1}{3!}\sum_{\gamma\beta\alpha} V_{\gamma\beta}^{CD}\,W_{\alpha\,DBA}$$
$$+ \sum_{\beta\alpha} \varepsilon_{\gamma\beta}\,\Lambda_{\alpha\,BA}^C \qquad (4.57)$$

$$\mathcal{D}_\alpha^C\,\overline{W}_{\dot{\beta}\dot{\alpha}}^{BA} = -\frac{1}{2}\sum_{\dot{\beta}\dot{\alpha}} \left(\mathrm{i}\,\mathcal{D}_{\alpha\dot{\beta}}\,\overline{W}_{\dot{\alpha}}^{CBA} + U_{\alpha\dot{\beta}\,D}^C\,\overline{W}_{\dot{\alpha}}^{DBA}\right)$$

$$\mathcal{D}_{\dot{\alpha}C}\,W_{\beta\alpha\,BA} = \frac{1}{2}\sum_{\beta\alpha} \left(\mathrm{i}\,\mathcal{D}_{\beta\dot{\alpha}}\,W_{\alpha\,CBA} - U_{\beta\dot{\alpha}\,C}^D\,W_{\alpha\,DBA}\right) \qquad (4.58)$$

$$\mathcal{D}_\beta^C\,U_{\alpha\dot{\alpha}\,A}^B = -\mathcal{D}_{\dot{\alpha}A}\,V_{\beta\alpha}^{CB} - 2\delta_A^C\,\overline{\Psi}_{\beta\alpha\dot{\alpha}}^B + \delta_A^B\,\overline{\Psi}_{\beta\alpha\dot{\alpha}}^C - W_{\beta\alpha\,AD}\,\overline{W}_{\dot{\alpha}}^{DCB}$$
$$+ \varepsilon_{\beta\alpha}\left(\mathcal{D}_{\dot{\alpha}A}\,S^{CB} - \overline{\Lambda}_{\dot{\alpha}A}^{CB} + 2\delta_A^C\,\overline{\Psi}_{\dot{\alpha}}^B + \delta_A^B\,\overline{\Psi}_{\dot{\alpha}}^C\right.$$
$$\left. -\frac{1}{3}\overline{V}_{\dot{\alpha}\dot{\beta}\,AD}\,\overline{W}^{\dot{\beta}DCB} + \overline{S}_{AD}\,\overline{W}_{\dot{\alpha}}^{DCB}\right)$$

$$\mathcal{D}_{\dot{\beta}C}\,U_{\alpha\dot{\alpha}\,A}^B = -\mathcal{D}_\alpha^B\,\overline{V}_{\dot{\beta}\dot{\alpha}\,CA} - 2\delta_C^B\,\Psi_{\dot{\beta}\dot{\alpha}\,\alpha A} + \delta_A^B\,\Psi_{\dot{\beta}\dot{\alpha}\,\alpha C} - \overline{W}_{\dot{\beta}\dot{\alpha}}^{BD}\,W_{\alpha\,DCA}$$
$$+ \varepsilon_{\dot{\beta}\dot{\alpha}}\left(\mathcal{D}_\alpha^B\,\overline{S}_{CA} - \Lambda_{\alpha\,CA}^B + 2\delta_C^B\,\Psi_{\alpha A} + \delta_A^B\,\Psi_{\alpha C}\right.$$
$$\left. -\frac{1}{3}V_{\alpha\beta}^{BD}\,W_{DCA}^\beta + S^{BD}\,W_{\alpha\,DCA}\right) \qquad (4.59)$$

---

[1]   Some of them are satisfied if one employs in addition the Ricci identity (2.23). It is, however, more convenient to keep all the conditions and to write them in the form of transformation laws under supersymmetry transformations.

$$\mathcal{D}_\gamma^C V_{\beta\alpha}^{BA} = V_{\gamma\beta\alpha}^{CBA} + \frac{1}{6} \sum_{\beta\alpha} \sum_{BA} \varepsilon_{\gamma\beta} \left(2 \mathcal{D}_\alpha^B S^{CA} + \frac{i}{2} \mathcal{D}_{\alpha\dot\alpha} \overline{W}^{\dot\alpha CBA} \right.$$

$$\left. - \frac{1}{2} U_{\alpha\dot\alpha\,D}^C \overline{W}^{\dot\alpha DBA} - 2 U_{\alpha\dot\alpha\,D}^B \overline{W}^{\dot\alpha DCA} \right)$$

$$\mathcal{D}_{\dot\gamma C} \overline{V}_{\dot\beta\dot\alpha\,BA} = \overline{V}_{\dot\gamma\dot\beta\dot\alpha\,CBA} + \frac{1}{6} \sum_{\dot\beta\dot\alpha} \sum_{BA} \varepsilon_{\dot\gamma\dot\beta} \left(2 \mathcal{D}_{\dot\alpha B} \overline{S}_{CA} - \frac{i}{2} \mathcal{D}_{\alpha\dot\alpha} W_{CBA}^\alpha \right.$$

$$\left. - \frac{1}{2} U_{\alpha\dot\alpha\,C}^D W_{DBA}^\alpha - 2 U_{\alpha\dot\alpha\,B}^D W_{DCA}^\alpha \right) \tag{4.60}$$

$$\sum_{CBA}^{\frown} \mathcal{D}_\alpha^C S^{BA} = \sum_{CBA}^{\frown} \mathcal{D}_{\dot\alpha C} \overline{S}_{BA} = 0 \tag{4.61}$$

5. Symmetries of the Constraints

Before we continue our analysis of the restrictions on the covariant superfields, we shall first derive the consequences of the constraints for the fundamental gauge potentials, the vielbein and the connection. That is, we shall not directly deal with $E_M{}^A$ and $\Phi_{MB}{}^A$. It is equivalent and much more convenient to consider their infinitesimal variations $H_B{}^A$ and $\Omega_{CB}{}^A$ (3.1–2).

First of all, the constraints are of course invariant under the general coordinate and structure group transformations (2.33–34). In terms of H and Ω, these equations read

$$H_B{}^A = \xi^C T_{CB}{}^A + \mathcal{D}_B \xi^A + \Lambda_B{}^A , \tag{5.1}$$

$$\Omega_{CB}{}^A = \xi^D R_{DCB}{}^A - \mathcal{D}_C \Lambda_B{}^A . \tag{5.2}$$

To find the additional invariances, we shall now simply go through the components of H and Ω listed in Table 2.

dim $-\frac{1}{2}$

The only component at dimension $-\frac{1}{2}$ is H_β^{Ba}. Using $\xi^{\underline{\alpha}}$ and (5.1), one can choose the gauge

$$H^B_{\beta\,\alpha\dot\alpha} = H^B_{(\beta\alpha)\dot\alpha} . \tag{5.3}$$

The superfield $H^B_{(\beta\alpha)\dot\alpha}$ contains (the variations of) all the covariant component fields of the conformal supergravity multiplet. At the next dimension we shall see, however, that it is not unconstrained.

dim 0

The dim-0 components are $H_b{}^a$ and $H_\beta{}^{\underline{\alpha}}$. Using $\Lambda_B{}^A$ and (5.1), we choose the gauge

$$H_{\beta\dot\beta\,\alpha\dot\alpha} = H_{(\beta\alpha)(\dot\beta\dot\alpha)} + 2\,\varepsilon_{\beta\alpha}\,\varepsilon_{\dot\beta\dot\alpha}\,H ,$$

$$H^{B\alpha}_{\alpha A} = H^{\dot\alpha B}_{A\dot\alpha} . \tag{5.4}$$

The conventional constraints (3.8) lead then via (3.5) to the algebraic equations

$$H_{\beta\dot\beta\,\alpha\dot\alpha} = 2\varepsilon_{\beta\alpha}\varepsilon_{\dot\beta\dot\alpha}\,H + \frac{i}{8N}\sum_{\beta\alpha}\sum_{\dot\beta\dot\alpha}(\mathcal{D}^A_{\dot\beta}\,H_{\dot\beta A\,\alpha\dot\alpha} + \mathcal{D}_{\dot\beta A}\,H^A_{\beta\,\alpha\dot\alpha}), \tag{5.5}$$

$$H^B_{\beta\,\alpha A} = \frac{1}{2}\varepsilon_{\beta\alpha}\delta^B_A\,H - \frac{i}{8}\mathcal{D}^{\dot\alpha}_A\,H^B_{\beta\,\alpha\dot\alpha}, \tag{5.6}$$

$$H^{BA}_{\beta\,\dot\alpha} = -\frac{i}{12}\left(\mathcal{D}^{\alpha B}\,H^A_{\alpha\,\beta\dot\alpha} - 2\,\mathcal{D}^{\alpha A}\,H^B_{\alpha\,\beta\dot\alpha} + 3\,\overline{W}^{BAC}_{\dot\beta}\,H^{\dot\beta}_{C\,\beta\dot\alpha}\right) \tag{5.7}$$

and the constraints (3.9) give the differential equations

$$\sum_{\gamma\beta\alpha}\sum_{CB}\mathcal{D}^C_\gamma\,H^B_{\beta\,\alpha\dot\alpha} = 0 , \tag{5.8}$$

$$\sum_{\gamma\alpha}\sum_{\dot\beta\dot\alpha} \mathrm{tl}\left(\mathcal{D}^C_\gamma\, H_{\dot\beta B\,\alpha\dot\alpha} + \mathcal{D}_{\dot\beta B}\, H^C_{\gamma\alpha\dot\alpha}\right) = 0\,, \tag{5.9}$$

where tl denotes the completely traceless part (see Appendix A). The superfield H remains unconstrained and parametrizes the super-Weyl transformations [15, 11, 12], which include the dilatations and the special supersymmetry transformations in space-time. In the remainder of this chapter we shall focus on these transformations, i.e., we shall assume that

$$H_{\underline\beta}{}^{a} = 0\,, \tag{5.10}$$

which implies[1]

$$H_{b}{}^{a} = -\delta^a_b\, H\,, \tag{5.11}$$

$$H_{\underline\beta}{}^{\underline\alpha} = -\frac{1}{2}\,\delta^{\underline\alpha}_{\underline\beta}\, H\,. \tag{5.12}$$

dim $\frac{1}{2}$

The dim-$\frac{1}{2}$ components $H_{b\dot\alpha}{}^{A}$ and $\Omega^C_{\gamma B}{}^{A}$ are completely fixed by the constraints (3.10). For super-Weyl transformations, one finds

$$H_{\beta\dot\beta\,\dot\alpha}{}^{A} = -\mathrm{i}\,\varepsilon_{\dot\beta\dot\alpha}\, \mathcal{D}^A_\beta\, H\,, \tag{5.13}$$

$$\Omega^C_{\gamma\beta\alpha} = \sum_{\beta\alpha}\varepsilon_{\gamma\beta}\, \mathcal{D}^C_\alpha\, H\,, \tag{5.14}$$

$$\Omega^C_{\gamma\dot\beta\dot\alpha} = 0\,, \tag{5.15}$$

$$\Omega^{CB}_{\gamma\ A} = 2\,\delta^C_A\, \mathcal{D}^B_\gamma\, H - \frac{1}{2}\,\delta^B_A\, \mathcal{D}^C_\gamma\, H\,. \tag{5.16}$$

dim 1

At dimension 1 the constraints (3.11) and (4.12) imply

$$\Omega_{cba} = \sum_{ba}\eta_{cb}\, \mathcal{D}_a\, H\,, \tag{5.17}$$

$$\Omega_c{}^{B}{}_{A} = 0\,. \tag{5.18}$$

Note that the U(1) connection is super-Weyl invariant for $N = 4$. This is a direct consequence of the superconformal algebra [3] (and of our choice of (4.12)).

---

[1] The factors are chosen such that the Weyl weight of a Weyl-covariant field is equal to its mass dimension.

Finally, we consider the transformation properties of the various superfields introduced in Chapter 4 under super-Weyl transformations. The basic Weyl-covariant superfield is

$$\overline{W}_{\dot\alpha}^{CBA} \ (N \geq 3), \quad \overline{W}_{\dot\beta\dot\alpha}^{BA} \ (N = 2), \quad \overline{W}_{\dot\gamma\dot\beta\dot\alpha}^{A} \ (N = 1). \tag{5.19}$$

From (3.5) one obtains

$$\delta \overline{W}_{\dot\alpha}^{CBA} = \frac{1}{2} H \overline{W}_{\dot\alpha}^{CBA}, \tag{5.20}$$

$$\delta \overline{W}_{\dot\beta\dot\alpha}^{BA} = H \overline{W}_{\dot\beta\dot\alpha}^{BA} - \frac{1}{2} \sum_{\dot\beta\dot\alpha} \overline{W}_{\dot\beta}^{BAC} \mathcal{D}_{\dot\alpha C} H, \tag{5.21}$$

$$\delta \overline{W}_{\dot\gamma\dot\beta\dot\alpha}^{A} = \frac{3}{2} H \overline{W}_{\dot\gamma\dot\beta\dot\alpha}^{A} + \frac{1}{3} \sum_{\dot\gamma\dot\beta\dot\alpha} \overline{W}_{\dot\gamma\dot\beta}^{AB} \mathcal{D}_{\dot\alpha B} H. \tag{5.22}$$

Similarly, the transformation laws of all the covariant derivatives of $\overline{W}$ may be computed. The only interesting result is

$$\delta \overline{M}_D^{CBA} = H \overline{M}_D^{CBA} + 2 \overline{W}_{\dot\alpha}^{CBA} \mathcal{D}_D^{\dot\alpha} H$$

$$- \frac{1}{2} \sum_{CBA} \delta_D^C \overline{W}_{\dot\alpha}^{BAE} \mathcal{D}_E^{\dot\alpha} H. \tag{5.23}$$

For $N = 4$, we find

$$\delta \overline{M}_C^{CBA} = H \overline{M}_C^{CBA}. \tag{5.24}$$

This means that $\overline{M}_C^{CBA}$ can be set equal to zero without breaking the super-Weyl invariance. Hence the $N = 4$ multiplet is still reducible. We shall come back to this at the end of Chapter 7.

The basic non-covariant superfields are

$$V_{\beta\alpha}^{BA}, \quad S^{BA}, \quad U_{\alpha\dot\alpha A}^{B} \tag{5.25}$$

with the transformation laws

$$\delta V_{\beta\alpha}^{BA} = H V_{\beta\alpha}^{BA} + \frac{1}{4} \sum_{\beta\alpha} [\mathcal{D}_\beta^B, \mathcal{D}_\alpha^A] H, \tag{5.26}$$

$$\delta S^{BA} = H S^{BA} - \frac{1}{4} [\mathcal{D}^{\alpha B}, \mathcal{D}_\alpha^A] H, \tag{5.27}$$

$$\delta U_{\alpha\dot\alpha A}^{B} = H U_{\alpha\dot\alpha A}^{B} - \frac{1}{2} [\mathcal{D}_\alpha^B, \mathcal{D}_{\dot\alpha A}] H. \tag{5.28}$$

They contain all the non-covariant component fields corresponding to the unconstrained superfield H.

6. Consistency of the Constraints

The constraints that we have imposed may be divided into two classes. First, there are the conventional constraints which imply algebraic equations for the vielbein and the connection. They are perfectly consistent. Second, there are the constraints (3.9) which are equivalent to the differential equations (5.8–9). In curved superspace these differential equations are non-linear, so we have to worry about their consistency.

Since "consistency" in this context is not very well defined, we illustrate by a simple example what might go wrong. Consider the covariant constraint

$$D_a V_b = 0 \tag{6.1}$$

in ordinary Riemannian geometry. From the chain of equations

$$0 = [D_a, D_b] V_c = -R_{abc}{}^d V_d \tag{6.2}$$

we conclude that either $V_a = 0$ or the space is flat. Hence (6.1) is in general inconsistent.

Guided by this example, it is easy to give a criterion for the consistency of constraints like (6.1). First define

$$X_{ab} = D_a V_b. \tag{6.3}$$

Next look for the constraint on X_{ab} which has the general solution (6.3). In our case this is

$$D_a X_{bc} - D_b X_{ac} + R_{abc}{}^d V_d = 0. \tag{6.4}$$

Then the constraint $X_{ab} = 0$ is consistent if (6.4) vanishes identically for $X_{ab} = 0$. Of course, this looks completely trivial, but in curved superspace it is not always possible to find equations like (6.2) "by inspection".

Now we turn to the constraints (5.8–9) and define

$$X^{(CB)}_{(\gamma\beta\alpha)\dot\alpha} = \frac{1}{12} \sum_{\gamma\beta\alpha} \widehat{\sum_{CB}} \mathcal{D}^C_\gamma H^B_{\beta\,\alpha\dot\alpha}, \tag{6.5}$$

$$Y^C_{(\gamma\alpha)(\dot\beta\dot\alpha)B} = \frac{1}{4} \sum_{\gamma\alpha} \sum_{\dot\beta\dot\alpha} \mathrm{tl}\left(\mathcal{D}^C_\gamma H_{\dot\beta B\,\alpha\dot\alpha} + \mathcal{D}_{\dot\beta B} H^C_{\gamma\,\alpha\dot\alpha}\right). \tag{6.6}$$

In order to find the constraints on X and Y implied by (6.5–6), we first consider the simplest approximation with only partial derivatives:

$$X^{(CB)}_{(\gamma\beta\alpha)\dot\alpha} = \frac{1}{12} \sum_{\gamma\beta\alpha} \widehat{\sum_{CB}} \partial^C_\gamma H^B_{\beta\,\alpha\dot\alpha}, \tag{6.7}$$

$$Y^C_{(\gamma\alpha)(\dot\beta\dot\alpha)B} = \frac{1}{4} \sum_{\gamma\alpha} \sum_{\dot\beta\dot\alpha} \mathrm{tl}\left(\partial^C_\gamma H_{\dot\beta B\,\alpha\dot\alpha} + \partial_{\dot\beta B} H^C_{\gamma\,\alpha\dot\alpha}\right). \tag{6.8}$$

The constraint on X corresponding to $\partial^{(A}_{(\alpha} \partial^{B)}_{\beta)} = 0$ is

$$\sum_{\delta\gamma\beta\alpha} \widehat{\sum_{DCB}} \partial^D_\delta X^{CB}_{\gamma\beta\alpha\,\dot\alpha} = 0. \tag{6.9}$$

23

There is no constraint corresponding to $\partial^{\alpha[A} \partial_\alpha^{B]} = 0$ since X is symmetric in CB. For Y we find

$$\sum_{\gamma\beta\alpha} \sum_{\dot\gamma\dot\beta\dot\alpha} \sum_{DC}^{\frown} \sum_{BA}^{\frown} \text{tl}\, \partial_\gamma^B \partial_{\dot\gamma D}\, Y^A_{\beta\alpha\,\dot\beta\dot\alpha\,C} = 0 \tag{6.10}$$

because of $\partial_{(\alpha}^{(A} \partial_{\beta)}^{B)} = 0$ and

$$\sum_{FED} \sum_{CBA} \text{tl}\, \partial^{\alpha C} \partial^{\beta B} \partial_F^{\dot\alpha} \partial_E^{\dot\beta}\, Y^A_{\beta\alpha\,\dot\beta\dot\alpha\,D} = 0 \tag{6.11}$$

because of $\partial^{\alpha[A} \partial_\alpha^{B]} = 0$. Finally, there is the mixed constraint

$$\sum_{\gamma\beta\alpha} \sum_{\dot\beta\dot\alpha} \sum_{CB}^{\frown} \text{tl}\, \big(\partial_{\dot\beta A}\, X^{CB}_{\gamma\beta\alpha\,\dot\alpha} + \partial_\gamma^C\, Y^B_{\beta\alpha\,\dot\beta\dot\alpha\,A} \big) = 0\,. \tag{6.12}$$

All other restrictions on X and Y are derivatives of the above four equations. Actually, (6.10) is already implied by (6.12), so we are left with the three independent constraints (6.9, 11, 12).

In the next step we extend these equations to the non-linear level. This gives

$$\sum_{\delta\gamma\beta\alpha} \sum_{DCB}^{\frown} \mathcal{D}_\delta^D X^{CB}_{\gamma\beta\alpha\,\dot\alpha} = 0\,, \tag{6.13}$$

$$\sum_{\gamma\beta\alpha} \sum_{\dot\beta\dot\alpha} \sum_{CB}^{\frown} \text{tl}\, \big(\mathcal{D}_{\dot\beta A}\, X^{CB}_{\gamma\beta\alpha\,\dot\alpha} + \mathcal{D}_\gamma^C\, Y^B_{\beta\alpha\,\dot\beta\dot\alpha\,A} \big) = 0\,, \tag{6.14}$$

$$\sum_{FED} \sum_{CBA} \text{tl}\, \big(\mathcal{D}^{\alpha C} \mathcal{D}^{\beta B} \mathcal{D}_F^{\dot\alpha} \mathcal{D}_E^{\dot\beta}\, Y^A_{\beta\alpha\,\dot\beta\dot\alpha\,D} + \mathcal{D}_F^{\dot\alpha} \mathcal{D}_E^{\dot\beta} \mathcal{D}^{\alpha C} \mathcal{D}^{\beta B}\, Y^A_{\beta\alpha\,\dot\beta\dot\alpha\,D}$$
$$+ \text{non-linear terms} \big) = 0\,. \tag{6.15}$$

Then we impose the original constraints (5.8–9)

$$X^{CB}_{\gamma\beta\alpha\,\dot\alpha} = Y^C_{\gamma\alpha\,\dot\beta\dot\alpha\,B} = 0 \tag{6.16}$$

and obtain

$$\sum_{FED} \sum_{CBA} \text{tl}\, (\text{non-linear terms})^{CBA}_{FED} = 0 \qquad (N \geq 6)\,. \tag{6.17}$$

Obviously this equation is identically satisfied for $N \leq 5$. For $N \geq 6$ it does not vanish, as we shall see in Chapter 8. In order to derive an inconsistency, however, it is more convenient to use the finite equation instead of its infinitesimal version. This is why we did not compute the non-linear terms in (6.15) and (6.17).

The conclusion of this chapter is that our constraints are consistent for $N \leq 5$. Since we shall impose only one more constraint for $N = 4$, this means that the $N \leq 3$ off-shell and $N = 5$ on-shell conformal supergravities are algebraically consistent.

7. Linearized Ricci Identities

In this chapter we are going to determine the conformal supergravity multiplet. In principle this could be done by computing the θ-expansions of the vielbein and the connection. It is, however, advantageous to keep the calculation as covariant as possible. Therefore we shall start from the superfields introduced in Chapter 4 and from the restrictions imposed by the Bianchi identities. Furthermore, we shall compute the covariant spinor derivatives of these superfields instead of their θ-expansions and we shall employ the Ricci identity (2.23) instead of the anticommutation relation for the θ-variables. In order to find the multiplet, it suffices to work in the linear approximation. Some non-linear results will be analyzed in the next chapter.

dim $\frac{3}{2}$

We start at dimension $\frac{3}{2}$ with the Ricci identities for $\overline{W}_{\dot{\alpha}}^{CBA}$. Using (2.23) and the solution of the Bianchi identities, one obtains from $\{D_{\beta}^{E}, D_{\alpha}^{D}\}\overline{W}_{\dot{\alpha}}^{CBA}$:

$$D_{\beta}^{E}\,\overline{N}_{\alpha\dot{\alpha}}^{DCBA} = i\,\overline{N}_{(\beta\alpha)\dot{\alpha}}^{[EDCBA]}. \tag{7.1}$$

Analogously, $\{D_{\alpha}^{E}, D_{\dot{\beta}D}\}\overline{W}_{\dot{\alpha}}^{CBA}$ implies

$$D_{\dot{\beta}E}\,\overline{N}_{\alpha\dot{\alpha}}^{DCBA} = i\varepsilon_{\dot{\beta}\dot{\alpha}}\,\overline{M}_{\alpha E}^{[DCBA]} - \frac{1}{6}\sum_{\dot{\beta}\dot{\alpha}}\sum_{DCBA}\delta_{E}^{D}\,\partial_{\alpha\dot{\beta}}\overline{W}_{\dot{\alpha}}^{CBA}, \tag{7.2}$$

$$D_{\alpha}^{E}\,\overline{M}_{D}^{CBA} = \overline{M}_{\alpha D}^{ECBA} + i\delta_{D}^{E}\,\partial_{\alpha\dot{\alpha}}\overline{W}^{\dot{\alpha}CBA} \tag{7.3}$$

and $\{D_{\dot{\gamma}E}, D_{\dot{\beta}D}\}\overline{W}_{\dot{\alpha}}^{CBA}$ gives

$$D_{\dot{\alpha}E}\,\overline{M}_{D}^{CBA} = \sum_{CBA}\left(\delta_{D}^{C}\,\overline{\Lambda}_{\dot{\alpha}E}^{BA} - 2\delta_{E}^{C}\,\overline{\Lambda}_{\dot{\alpha}D}^{BA}\right). \tag{7.4}$$

PROOF: As an example we show how the first result (7.1) can be derived. The linearized Ricci identity is

$$\{D_{\beta}^{E}, D_{\alpha}^{D}\}\overline{W}_{\dot{\alpha}}^{CBA} = 0.$$

Using (4.15), this can be written as

$$D_{\beta}^{E}\,\overline{N}_{\alpha\dot{\alpha}}^{DCBA} + D_{\alpha}^{D}\,\overline{N}_{\beta\dot{\alpha}}^{ECBA} = 0. \tag{a}$$

Next we decompose

$$D_{\beta}^{E}\,\overline{N}_{\alpha\dot{\alpha}}^{DCBA} = i\,\overline{N}_{(\beta\alpha)\dot{\alpha}}^{E[DCBA]} + i\varepsilon_{\beta\alpha}\,\overline{N}_{\dot{\alpha}}^{E[DCBA]}.$$

Equation (a) then splits into

$$\sum_{ED}^{\wedge}\overline{N}_{(\beta\alpha)\dot{\alpha}}^{E[DCBA]} = 0, \tag{b}$$

$$\sum_{ED} \overline{N}_{\dot\alpha}^{E[DCBA]} = 0 \,. \qquad\qquad (c)$$

Inserting (b) into the decomposition

$$\overline{N}_{(\beta\alpha)\dot\alpha}^{E[DCBA]} = \overline{N}_{(\beta\alpha)\dot\alpha}^{[EDCBA]} + \frac{1}{30} \sum_{DCBA} \widehat{\sum_{ED}} \overline{N}_{(\beta\alpha)\dot\alpha}^{E[DCBA]}$$

gives

$$\overline{N}_{(\beta\alpha)\dot\alpha}^{E[DCBA]} = \overline{N}_{(\beta\alpha)\dot\alpha}^{[EDCBA]} \,.$$

From (c) and the identity

$$\overline{N}_{\dot\alpha}^{E[DCBA]} = \frac{1}{4!} \sum_{DCBA} \left(\sum_{ED} \overline{N}_{\dot\alpha}^{E[DCBA]} - \overline{N}_{\dot\alpha}^{D[CBAE]} \right)$$

we obtain

$$\overline{N}_{\dot\alpha}^{E[DCBA]} = 0 \,.$$

The final result is then given by (7.1).

dim 2

At dimension 2 we have to consider quite a lot of Ricci identities:

$$\{D_\beta^D, D_\alpha^C\} \overline{W}_{\dot\beta\dot\alpha}^{BA} \,, \quad \{D_\alpha^D, D_{\dot\gamma C}\} \overline{W}_{\dot\beta\dot\alpha}^{BA} \,, \quad \{D_\gamma^D, D_\beta^C\} U_{\alpha\dot\alpha\,A}^{B} \,^*,$$

$$\{D_\beta^D, D_{\dot\beta C}\} U_{\alpha\dot\alpha\,A}^{B} \,, \quad \{D_{\dot\delta D}, D_{\dot\gamma C}\} \overline{W}_{\dot\beta\dot\alpha}^{BA} \,^*,$$

$$\{D_\delta^D, D_\gamma^C\} V_{\beta\alpha}^{BA} \,, \quad \widehat{\sum_{DCBA}} \{D_\beta^D, D_\alpha^C\} S^{BA} \,^*,$$

$$\{D_\gamma^F, D_\beta^E\} \overline{N}_{\alpha\dot\alpha}^{DCBA} \,, \quad \{D_\beta^F, D_{\dot\beta E}\} \overline{N}_{\alpha\dot\alpha}^{DCBA} \,, \quad \{D_\beta^F, D_\alpha^E\} \overline{M}_D^{CBA} \,^*,$$

$$\{D_\alpha^F, D_{\dot\alpha E}\} \overline{M}_D^{CBA} \,, \quad \{D_{\dot\gamma F}, D_{\dot\beta E}\} \overline{N}_{\alpha\dot\alpha}^{DCBA} \,^*, \quad \{D_{\dot\beta F}, D_{\dot\alpha E}\} \overline{M}_D^{CBA} \,. \qquad (7.5)$$

The identities marked by * are satisfied on account of the preceding ones. Below we list the results following from the remaining Ricci identities. (Results from the same identity are numbered together.)

$$\sum_{\dot\beta\dot\alpha} \partial^\alpha_{\dot\beta} \overline{N}_{\alpha\dot\alpha}^{DCBA} = 0 \,, \qquad\qquad (7.6)$$

$$D_\alpha^D \overline{\Lambda}_{\dot\alpha C}^{BA} = \frac{i}{2} \partial_{\alpha\dot\alpha} \overline{M}_C^{DBA} - i\delta_C^D \partial_\alpha{}^{\dot\beta} \overline{W}_{\dot\beta\dot\alpha}^{BA} + \frac{i}{3} \sum_{BA} \delta_C^B \partial_\alpha{}^{\dot\beta} \overline{W}_{\dot\beta\dot\alpha}^{DA} \,, \qquad (7.7)$$

$$D_{\dot\beta D}\, \overline{\Lambda}^{BA}_{\dot\alpha C} = \varepsilon_{\dot\beta\dot\alpha}\, P^{[BA]}_{[DC]} + \sum_{BA}\left(\delta^B_D\, \overline{\rho}^A_{\dot\beta\dot\alpha\, C} - \frac{1}{3}\,\delta^B_C\,\overline{\rho}^A_{\dot\beta\dot\alpha\, D}\right),$$

$$D_{\dot\beta D}\, D_{\dot\alpha C}\, S^{BA} = S^{(BA)}_{(\dot\beta\dot\alpha)[DC]} + \varepsilon_{\dot\beta\dot\alpha}\left(S^{(BA)}_{(DC)} - \frac{i}{4}\sum_{DC}\sum_{BA}\delta^B_D\,\widehat{\partial}^{\gamma\dot\gamma}\,U^A_{\gamma\dot\gamma\, C}\right),$$

$$D_{\dot\beta D}\, D_{\dot\alpha C}\, V^{BA}_{\beta\alpha} = V^{[BA]}_{(\beta\alpha)(\dot\beta\dot\alpha)[DC]} + \frac{i}{4}\sum_{\beta\alpha}\sum_{\dot\beta\dot\alpha}\sum_{DC}\sum_{BA}\delta^B_D\,\partial_{\beta\dot\beta}\,U^A_{\alpha\dot\alpha\, C}$$

$$- \varepsilon_{\dot\beta\dot\alpha}\,\overline{S}^{BA}_{\beta\alpha\, DC} + \frac{1}{2}\varepsilon_{\dot\beta\dot\alpha}\sum_{\beta\alpha}\sum_{DC}\sum_{BA}\delta^B_D\left(\rho^A_{\beta\alpha\, C} + i\partial_\beta{}^{\dot\gamma}\,U^A_{\alpha\dot\gamma\, C}\right),$$

$$P^{BA}_{DC} = \overline{P}^{BA}_{DC}, \qquad S^{BA}_{DC} = \overline{S}^{BA}_{DC}, \qquad V^{BA}_{\beta\alpha\,\dot\beta\dot\alpha\, DC} = \overline{V}^{BA}_{\beta\alpha\,\dot\beta\dot\alpha\, DC}, \tag{7.8}$$

$$D^D_\delta\, V^{CBA}_{\gamma\beta\alpha} = V^{[DCBA]}_{(\delta\gamma\beta\alpha)} - \frac{1}{24}\sum_{\gamma\beta\alpha}\sum_{CBA}\varepsilon_{\delta\gamma}\left(D^C_\beta\, D^B_\alpha\, S^{AD} - \frac{1}{6}\partial_\beta{}^{\dot\alpha}\,\overline{N}^{DCBA}_{\alpha\dot\alpha}\right), \tag{7.9}$$

$$D^F_\gamma\, \overline{N}^{EDCBA}_{\beta\alpha\,\dot\alpha} = \overline{N}^{[FEDCBA]}_{(\gamma\beta\alpha)\dot\alpha}, \tag{7.10}$$

$$D_{\dot\beta F}\, \overline{N}^{EDCBA}_{\beta\alpha\,\dot\alpha} = \varepsilon_{\dot\beta\dot\alpha}\,\overline{M}^{[EDCBA]}_{(\beta\alpha)F} - \frac{1}{48}\sum_{\beta\alpha}\sum_{\dot\beta\dot\alpha}\sum_{EDCBA}\delta^E_F\,\partial_{\beta\dot\beta}\,\overline{N}^{DCBA}_{\alpha\dot\alpha},$$

$$D^F_\beta\, \overline{M}^{DCBA}_{\alpha E} = -\overline{M}^{FDCBA}_{\beta\alpha E} - \delta^F_E\,\partial_\beta{}^{\dot\alpha}\,\overline{N}^{DCBA}_{\alpha\dot\alpha}, \tag{7.11}$$

$$D_{\dot\alpha F}\, \overline{M}^{DCBA}_{\alpha E} = \frac{i}{6}\sum_{DCBA}\left(\delta^D_E\,\partial_{\alpha\dot\alpha}\,\overline{M}^{CBA}_F - 2\delta^D_F\,\partial_{\alpha\dot\alpha}\,\overline{M}^{CBA}_E\right.$$

$$\left. + 6\,\delta^D_F\,\delta^C_E\,\partial_\alpha{}^{\dot\beta}\,\overline{W}^{BA}_{\dot\beta\dot\alpha}\right), \tag{7.12}$$

$$\sum_{FED}\sum_{CBA}\delta^C_F\, P^{BA}_{ED} = 0. \tag{7.13}$$

The last equation is equivalent to

$$P^{AB}_{AB} = 0 \qquad (N=3),$$

$$P^{AC}_{BC} = 0 \qquad (N=4),$$

$$P^{BA}_{DC} = 0 \qquad (N>4). \tag{7.14}$$

Second Bianchi Identity

At the end of Chapter 4 we have cited a theorem by Dragon showing that the second Bianchi identity is a consequence of the first Bianchi identity and the Ricci identity. At dimensions $\frac{5}{2}$ and 3, however, it is more convenient to start from the Bianchi identities and to show that certain Ricci identities are then satisfied.

The linearized second Bianchi identity (2.27) reads at dimensions $\frac{5}{2}$ and 3:

$$D_{\underline{\varepsilon}} R_{dc\,B}{}^{A} = \sum_{dc} \partial_d R_{\underline{\varepsilon}\,c\,B}{}^{A}\,, \tag{7.15}$$

$$\sum_{edc} \partial_e R_{dc\,B}{}^{A} = 0\,. \tag{7.16}$$

From (7.15) one obtains

$$D_{\gamma}^{A} P_{\beta\alpha\,\dot\beta\dot\alpha} = \frac{i}{4} \sum_{\beta\alpha} \sum_{\dot\beta\dot\alpha} \left(2\,\partial_{\gamma\dot\beta}\,\overline{\Psi}_{\beta\alpha\dot\alpha}^{A} - \partial_{\beta\dot\beta}\,\overline{\Psi}_{\gamma\alpha\dot\alpha}^{A} + 3\,\varepsilon_{\gamma\beta}\,\partial_{\alpha\dot\beta}\,\overline{\Psi}_{\dot\alpha}^{A} \right), \tag{7.17}$$

$$D_{\varepsilon}^{A} W_{\delta\gamma\beta\alpha} = \frac{i}{12} \sum_{\delta\gamma\beta\alpha} \varepsilon_{\varepsilon\delta}\,\partial_{\gamma}{}^{\dot\alpha}\,\overline{\Psi}_{\beta\alpha\dot\alpha}^{A}\,,$$

$$D_{\alpha}^{A} \overline{W}_{\dot\delta\dot\gamma\dot\beta\dot\alpha} = \frac{i}{12} \sum_{\dot\delta\dot\gamma\dot\beta\dot\alpha} \partial_{\alpha\dot\delta}\,\overline{W}_{\dot\gamma\dot\beta\dot\alpha}^{A}\,, \tag{7.18}$$

$$D_{\alpha}^{A} R = -12i\,\partial_{\alpha}{}^{\dot\alpha}\,\overline{\Psi}_{\dot\alpha}^{A}\,, \tag{7.19}$$

$$D_{\gamma}^{C} \rho_{\beta\alpha\,A}^{B} = \frac{i}{3!} \sum_{\gamma\beta\alpha} \partial_{\gamma}{}^{\dot\alpha} \left(2\delta_{A}^{C}\,\overline{\Psi}_{\beta\alpha\dot\alpha}^{B} - \frac{1}{2}\delta_{A}^{B}\,\overline{\Psi}_{\beta\alpha\dot\alpha}^{C} \right) + i \sum_{\beta\alpha} \varepsilon_{\gamma\beta}\,\partial_{\alpha}{}^{\dot\alpha}\,\overline{\Lambda}_{\dot\alpha A}^{CB}\,,$$

$$D_{\alpha}^{C} \overline{\rho}_{\dot\beta\dot\alpha\,A}^{B} = -i\,\partial_{\alpha}{}^{\dot\gamma} \left(2\delta_{A}^{C}\,\overline{W}_{\dot\gamma\dot\beta\dot\alpha}^{B} - \frac{1}{2}\delta_{A}^{B}\,\overline{W}_{\dot\gamma\dot\beta\dot\alpha}^{C} \right) - i \sum_{\dot\beta\dot\alpha} \partial_{\alpha\dot\beta}\,\overline{\Lambda}_{\dot\alpha A}^{CB} \tag{7.20}$$

and (7.16) yields

$$\sum_{\gamma\beta\alpha} \left(\partial^{\delta}{}_{\dot\alpha} W_{\delta\gamma\beta\alpha} - \partial_{\gamma}{}^{\dot\beta} P_{\beta\alpha\,\dot\beta\dot\alpha} \right) = 0\,, \tag{7.21}$$

$$\partial^{\beta\dot\beta} P_{\beta\alpha\,\dot\beta\dot\alpha} + \frac{1}{4}\partial_{\alpha\dot\alpha} R = 0\,, \tag{7.22}$$

$$\partial^{\beta}{}_{\dot\alpha} \rho_{\beta\alpha\,A}^{B} + \partial_{\alpha}{}^{\dot\beta} \overline{\rho}_{\dot\beta\dot\alpha\,A}^{B} = 0\,. \tag{7.23}$$

The corresponding non-linear equations are given in Appendix E.

It is now easy to show that the Ricci identities for the highest-dimensional torsion and curvature components do not contain any new information. Note that the following theorem holds on the non-linear level and without any constraints.

28

THEOREM: The Ricci identities for $T_{cb}{}^\alpha$ and $R_{dc\,B}{}^A$ follow from the Bianchi identities and the Ricci identities for lower-dimensional components of the torsion and the curvature.

PROOF: Applying $\mathcal{D}$ to the first Bianchi identity (2.21) and using the second Bianchi identity (2.15) gives $\mathcal{D}\mathcal{D}T^A = T^B R_B{}^A$ or

$$E^B E^C \left(\mathcal{D}\mathcal{D}T_{CB}{}^A + 2 R_C{}^D T_{DB}{}^A - R_D{}^A T_{CB}{}^D \right) = 0 \,. \tag{7.24}$$

Next we define

$$T_{\mathcal{E}DCB}{}^A = [\mathcal{D}_{\mathcal{E}}, \mathcal{D}_D\} T_{CB}{}^A + T_{\mathcal{E}D}{}^{\mathcal{F}} \mathcal{D}_{\mathcal{F}} T_{CB}{}^A$$
$$+ \sum_{CB} R_{\mathcal{E}DC}{}^{\mathcal{F}} T_{\mathcal{F}B}{}^A - R_{\mathcal{E}D\mathcal{F}}{}^A T_{CB}{}^{\mathcal{F}} \,. \tag{7.25}$$

Equation (7.24) is then equivalent to

$$\sum_{\mathcal{E}DCB} T_{\mathcal{E}DCB}{}^A = 0 \,, \tag{7.26}$$

which, in turn, implies

$$T_{\underline{\varepsilon}\,\underline{\delta}\,cb}{}^\alpha = -\sum_{\underline{\varepsilon}\,\underline{\delta}} \sum_{cb} T_{\underline{\varepsilon}\,cb\,\underline{\delta}}{}^\alpha - T_{cb\,\underline{\varepsilon}\,\underline{\delta}}{}^\alpha \,,$$

$$\sum_{dcb} T_{\underline{\varepsilon}\,dcb}{}^\alpha = \sum_{dcb} T_{dcb\,\underline{\varepsilon}}{}^\alpha \,,$$

$$\sum_{edcb} T_{edcb}{}^\alpha = 0 \,. \tag{7.27}$$

The left-hand sides of these equations are the non-trivial Ricci identities for $T_{cb}{}^\alpha$, while the right-hand sides are trivial Ricci identities for lower-dimensional torsion components. The proof for $R_{dc\,B}{}^A$ is quite similar and will therefore be omitted.

dim $\frac{5}{2}$

We now continue our analysis of the linearized Ricci identities, but in the following we shall consider only those identities that give Weyl-covariant results. Furthermore, we shall omit all terms which are antisymmetric in more than five indices. The reason is that the non-linear version of (7.13) is inconsistent for $N \geq 6$, as we shall see in the next chapter.

At dimension $\frac{5}{2}$, the spinor derivatives of (7.6) yield

$$\sum_{\dot\beta\dot\alpha} \partial^\beta{}_{\dot\beta} \overline{N}^{EDCBA}_{\beta\alpha\,\dot\alpha} = 0 \,, \tag{7.28}$$

$$\partial^\alpha{}_{\dot\alpha} \overline{M}^{DCBA}_{\alpha E} = \frac{i}{12} \sum_{DCBA} \delta^D_E \, \partial^{\beta\dot\beta} \partial_{\beta\dot\beta} \overline{W}^{CBA}_{\dot\alpha} \,. \tag{7.29}$$

29

From $\{D_\alpha^E, D_{\dot\beta D}\}\,\overline\Lambda_{\dot\alpha C}^{BA}$ one obtains

$$D_\alpha^E P_{DC}^{BA} = \mathrm{i} \sum_{DC}\sum_{BA} \partial_\alpha{}^{\dot\alpha}\left(\delta_D^B\,\overline\Lambda_{\dot\alpha C}^{EA} - \delta_D^E\,\overline\Lambda_{\dot\alpha C}^{BA}\right). \tag{7.30}$$

Because of (7.14), this implies

$$\partial_\alpha{}^{\dot\alpha}\,\overline\Lambda_{\dot\alpha B}^{BA} = 0 \qquad (N = 4),$$
$$\partial_\alpha{}^{\dot\alpha}\,\overline\Lambda_{\dot\alpha C}^{BA} = 0 \qquad (N > 4). \tag{7.31}$$

Finally, $\{D_\gamma^G, D_{\dot\beta F}\}\,\overline N_{\beta\alpha\dot\alpha}^{EDCBA}$ gives

$$D_\gamma^G\,\overline M_{\beta\alpha F}^{EDCBA} = -\mathrm{i}\,\delta_F^G\,\partial_\gamma{}^{\dot\alpha}\,\overline N_{\beta\alpha\dot\alpha}^{EDCBA} \tag{7.32}$$

and $\{D_\beta^G, D_{\dot\alpha F}\}\,\overline M_{\alpha E}^{DCBA}$ leads to

$$D_{\dot\alpha G}\,\overline M_{\beta\alpha F}^{EDCBA} = \frac{\mathrm{i}}{48} \sum_{\beta\alpha}\sum_{EDCBA} \partial_{\beta\dot\alpha}\left(2\,\delta_G^E\,\overline M_{\alpha F}^{DCBA} - \delta_F^E\,\overline M_{\alpha G}^{DCBA}\right.$$
$$\left. + 4\mathrm{i}\,\delta_G^E\,\delta_F^D\,\partial_{\alpha\dot\beta}\,\overline W^{\dot\beta CBA}\right). \tag{7.33}$$

dim 3

At dimension 3 we apply $D_{\dot\gamma F}$ to (7.28),

$$\partial^\beta{}_{\dot\alpha}\,\overline M_{\beta\alpha F}^{EDCBA} = -\frac{1}{48} \sum_{EDCBA} \delta_F^E\,\partial^{\beta\dot\beta}\,\partial_{\beta\dot\beta}\,\overline N_{\alpha\dot\alpha}^{DCBA}, \tag{7.34}$$

and $D_{\dot\beta F}$ to (7.29),

$$\partial^{\alpha\dot\alpha}\,\partial_{\alpha\dot\alpha}\,\overline M_C^{CBA} = 0 \qquad (N = 4),$$
$$\partial^{\alpha\dot\alpha}\,\partial_{\alpha\dot\alpha}\,\overline M_D^{CBA} = 0 \qquad (N > 4). \tag{7.35}$$

Similarly, the spinor derivatives of (7.31) give

$$\partial_\alpha{}^{\dot\alpha}\,\partial_\beta{}^{\dot\beta}\,\overline W_{\dot\beta\dot\alpha}^{BA} = 0 \qquad (N > 4) \tag{7.36}$$

and

$$\partial_\alpha{}^{\dot\beta}\,\overline\rho_{\dot\beta\dot\alpha A}^A = 0 \qquad (N = 4),$$
$$\partial_\alpha{}^{\dot\beta}\,\overline\rho_{\dot\beta\dot\alpha A}^B = 0 \qquad (N > 4). \tag{7.37}$$

dim > 3

The spinor derivatives of (7.36) are at dimension $\frac{7}{2}$

$$\partial^{\beta\dot\beta} \partial_{\beta\dot\beta} \partial_{\alpha\dot\alpha} \overline{W}^{\dot\alpha CBA} = 0 \qquad (N > 4), \qquad (7.38)$$

$$\partial_\alpha{}^{\dot\beta} \partial_\beta{}^{\dot\gamma} \overline{W}^A_{\dot\gamma\dot\beta\dot\alpha} = 0 \qquad (N > 4) \qquad (7.39)$$

and at dimension 4

$$\partial^{\beta\dot\beta} \partial_{\beta\dot\beta} \partial^{\alpha\dot\alpha} \overline{N}^{DCBA}_{\alpha\dot\alpha} = 0 \qquad (N > 4), \qquad (7.40)$$

$$\partial_\alpha{}^{\dot\gamma} \partial_\beta{}^{\dot\delta} \overline{W}_{\dot\delta\dot\gamma\dot\beta\dot\alpha} = 0 \qquad (N > 4). \qquad (7.41)$$

The $\theta = \overline{\theta} = 0$ components of (7.28–29), (7.31), and (7.34–41) are conformal field equations. The last one is the field equation for the graviton.

These are all the Weyl-covariant equations that follow from the linearized Ricci identities for $N \leq 5$. Some of them were already found in Ref. [9]. An immediate conclusion is that $N > 4$ conformal supergravity has no off-shell formulation in conventional extended superspace.

N = 4

Before we summarize our results, we briefly discuss the case $N = 4$. In this case the U(1) sector is on-shell, whereas the rest of the multiplet is off-shell. Hence the $N = 4$ multiplet is fully reducible and the on-shell part may be eliminated by the constraint [11,12]

$$\overline{M}^{CBA}_C = 0. \qquad (7.42)$$

Equation (5.24) shows that this constraint is Weyl-covariant even on the non-linear level. The consequences of (7.42) are at dimension $\frac{3}{2}$

$$\overline{M}^{DCBA}_{\alpha E} = \frac{i}{6} \sum_{DCBA} \delta^D_E \partial_{\alpha\dot\alpha} \overline{W}^{\dot\alpha CBA}, \qquad (7.43)$$

$$\overline{\Lambda}^{BA}_{\dot\alpha B} = 0 \qquad (7.44)$$

and at dimension 2

$$\sum_{\beta\alpha} \partial_\beta{}^{\dot\alpha} \overline{N}^{DCBA}_{\alpha\dot\alpha} = 0, \qquad (7.45)$$

$$\overline{\rho}^A_{\dot\beta\dot\alpha A} = 0. \qquad (7.46)$$

Moreover, the Bianchi identities and the equations of motion for the above superfields vanish identically. Equations (7.6) and (7.45) can be solved by

$$\overline{N}^{DCBA}_{\alpha\dot\alpha} = \partial_{\alpha\dot\alpha} \overline{W}^{[DCBA]}, \qquad (7.47)$$

where

$$D_\alpha^E \, \overline{W}^{DCBA} = 0 \,, \tag{7.48}$$

$$D_{\dot\alpha E} \, \overline{W}^{DCBA} = -\frac{1}{3} \sum_{DCBA} \delta_E^D \, \overline{W}_{\dot\alpha}^{CBA} \,. \tag{7.49}$$

Summary

In Table 3 we have listed the Weyl-covariant superfields for $N \leq 5$. All of them are completely symmetric in the spinor indices and antisymmetric in the internal indices. The arrows indicate their transformation properties under supersymmetry transformations.

dim	superfields
$\frac{1}{2}$	$\overline{W}_{\dot\alpha}^{CBA}$
	$\swarrow \qquad \downarrow \qquad \searrow$
1	$\overline{W}_{\dot\beta\dot\alpha}^{BA} \qquad \overline{M}_D^{CBA} \qquad \overline{N}_{\alpha\dot\alpha}^{DCBA}$
	$\swarrow \; \downarrow \; \swarrow \quad \downarrow \quad \swarrow \quad \downarrow$
$\frac{3}{2}$	$\overline{W}_{\dot\gamma\dot\beta\dot\alpha}^{A} \quad \overline{\Lambda}_{\dot\alpha C}^{BA} \quad \overline{M}_{\alpha E}^{DCBA} \quad \overline{N}_{\beta\alpha\dot\alpha}^{EDCBA}$
	$\swarrow \; \downarrow \; \swarrow \; \downarrow \qquad \downarrow \; \swarrow$
2	$\overline{W}_{\dot\delta\dot\gamma\dot\beta\dot\alpha} \quad \overline{\rho}_{\dot\beta\dot\alpha A}^{B} \quad P_{DC}^{BA} \quad \overline{M}_{\beta\alpha F}^{EDCBA}$

Table 3. Weyl-covariant superfields for $N \leq 5$

The $\theta = \overline\theta = 0$ components of these superfields determine the conformal supergravity multiplet. For instance, $\overline{W}_{\dot\delta\dot\gamma\dot\beta\dot\alpha}\big|_{\theta=\overline\theta=0}$ is the Weyl tensor of the graviton (up to non-linear terms), $\overline{W}_{\dot\gamma\dot\beta\dot\alpha}^{A}\big|$ is the Weyl tensor of the gravitinos, $\overline{\rho}_{\dot\beta\dot\alpha A}^{B}\big|$ is the U(N) field strength, $P_{DC}^{BA}\big|$ is an auxiliary field, and so on. For $N = 4$, $\overline{N}_{\alpha\dot\alpha}^{DCBA}\big|$ is the field strength of a complex scalar field, but this is no longer the case for $N = 5$. $\overline{M}_{\beta\alpha F}^{EDCBA}\big|$ appears only for $N = 5$ and describes five vector fields in the fundamental representation of U(5) (in the linear approximation).[1] Since the square of this field cannot be U(5) invariant, there exists no invariant action for $N = 5$.

The counting of the degrees of freedom in $N \leq 4$ conformal supergravity is explained in detail in Refs. [16, 17, 4] and will not be repeated here. For $N = 5$, $\overline{N}_{\alpha\dot\alpha}\big|$ counts like a massless complex scalar plus a massive complex scalar plus a massless real vector with a total of 6 degrees of freedom. $\overline{N}_{\beta\alpha\dot\alpha}\big|$ is a non-gauge spin-$\frac{3}{2}$ field with 6 on-shell

---

[1] These gauge potentials do not exist in the non-linear theory. This is another reason why there is no action for $N = 5$.

states. Multiplying the degrees of freedom of each field by the dimension of the U(5) representation and adding up, one finds 256 bosonic and 256 fermionic states. Note, however, that this equality has nothing to do with the non-linear consistency of the theory. It follows already from the linear approximation which is consistent for any N.

The $N \leq 4$ conformal supergravity multiplets are once more listed below. They are written in the form: (higher-derivative fields | ordinary fields | auxiliary fields).

$$N = 1: \qquad \left(e_m{}^a,\ \psi_m{}^\alpha \mid v_m \right) \tag{7.50}$$

$$N = 2: \qquad \left(e_m{}^a,\ \psi_{mA}{}^\alpha \mid v_{mA}^B,\ \overline{\Lambda}_{\dot\alpha C}^{BA},\ \overline{W}_{\dot\beta\dot\alpha}^{BA} \mid P_{DC}^{BA} \right) \tag{7.51}$$

$$N = 3: \qquad \left(e_m{}^a,\ \psi_{mA}{}^\alpha,\ \overline{W}_{\dot\alpha}^{CBA} \mid v_{mA}^B,\ \overline{\Lambda}_{\dot\alpha C}^{BA},\ \overline{W}_{\dot\beta\dot\alpha}^{BA},\ \overline{M}_D^{CBA} \mid P_{DC}^{BA} \right) \tag{7.52}$$

$$N = 4: \qquad \left(e_m{}^a,\ \psi_{mA}{}^\alpha,\ \overline{W}_{\dot\alpha}^{CBA},\ W_i \mid v_{mA}^B,\ \overline{\Lambda}_{\dot\alpha C}^{BA},\ \overline{W}_{\dot\beta\dot\alpha}^{BA},\ \overline{M}_D^{CBA} \mid P_{DC}^{BA} \right) \tag{7.53}$$

$E_m{}^a| = e_m{}^a$ is the graviton, $E_m{}^\alpha_A| = \frac{1}{2}\psi_{mA}{}^\alpha$ are the gravitinos, and $\Phi_m{}^B{}_A| = i\,v_{mA}^B$ are the (S)U(N) gauge fields. The scalar fields W_i are the non-linear analogues of $\overline{W}^{DCBA}$. They will be described in the next chapter. All fields in (7.50–53) depend only on x^m.

8. Non-Linear Ricci Identities

In Appendix D we have specified the solution of the non-linear Ricci identities up to dimension 2, i.e., the non-linear versions of (7.1–4) and (7.6–13). We shall need these equations to show that $N = 4$ off-shell conformal supergravity is algebraically consistent and that $N \geq 6$ conformal supergravity is not.

$N = 4$

The algebraic consistency of $N = 4$ off-shell conformal supergravity can be understood through the following simple argument. Consider the reducible $N = 4$ multiplet which splits into an off-shell and an on-shell part. In Chapter 6 we have shown that it is algebraically consistent. Therefore it should be possible to express the off-shell part in terms of an unconstrained prepotential Π^{BA}_{DC} of dimension -6. The multiplet defined by Π is obviously consistent and corresponds with the one defined by the constraint (7.42).

Of course, this argument is not very rigorous, so we have nevertheless computed the non-linear versions of (7.42–46). First of all, we note that the constraint (7.42) remains unchanged. The only possible non-linear term on the right-hand side would be the square of $W^{\alpha}_{ABC} = \varepsilon_{ABCD} W^{\alpha D}$, but $W^{\alpha A} W^B_{\alpha}$ is symmetric in AB. The consequences of this constraint can then be found from the equations given in Appendix D. After a tedious calculation one obtains

$$
\overline{M}^{DCBA}_{\alpha D} = \frac{1}{3!} \sum_{CBA} \left(i \, \mathcal{D}_{\alpha\dot\alpha} \overline{W}^{\dot\alpha CBA} + U^D_{\alpha\dot\alpha D} \overline{W}^{\dot\alpha CBA} - 3 \, U^C_{\alpha\dot\alpha D} \overline{W}^{\dot\alpha DBA} \right.
$$
$$
\left. + \frac{1}{6} W_{\alpha DEF} \overline{W}^{DEF}_{\dot\alpha} \overline{W}^{\dot\alpha CBA} \right), \tag{8.1}
$$

$$
\overline{\Lambda}^{BA}_{\dot\alpha B} = \frac{1}{12} \left(2 \overline{V}_{\dot\alpha\dot\beta BC} \overline{W}^{\dot\beta CBA} + M^A_{BCD} \overline{W}^{DCB}_{\dot\alpha} + i \, W^{\alpha}_{BCD} \overline{N}^{DCBA}_{\alpha\dot\alpha} \right), \tag{8.2}
$$

$$
\sum_{\beta\alpha} \sum_{DCBA} \left(\mathcal{D}_{\beta}{}^{\dot\alpha} \overline{N}^{DCBA}_{\alpha\dot\alpha} - 8 \, \overline{\Psi}^D_{\beta\alpha\dot\alpha} \overline{W}^{\dot\alpha CBA} \right.
$$
$$
\left. - \frac{i}{6} W_{\beta EFG} \overline{W}^{\dot\alpha EFG} \overline{N}^{DCBA}_{\alpha\dot\alpha} \right) = 0, \tag{8.3}
$$

$$
\overline{\rho}^A_{\dot\beta\dot\alpha A} = \frac{1}{12} \sum_{\dot\beta\dot\alpha} \left[\frac{1}{8} N^{\alpha}{}_{\dot\beta ABCD} \overline{N}^{ABCD}_{\alpha\dot\alpha} - i \, \mathcal{D}_{\alpha\dot\beta} \left(W^{\alpha}_{ABC} \overline{W}^{ABC}_{\dot\alpha} \right) \right]. \tag{8.4}
$$

The simplicity of these results suggests that there is an underlying geometric structure. Indeed, the above equations can also be derived by extending the local $U(1)$ symmetry to $SU(1,1)$ and setting the $SU(1,1)$ curvature equal to zero [11,12]. The $SU(1,1)$ connection then becomes the Maurer-Cartan form of a superfield $\mathcal{W} \in SU(1,1)$. It is convenient to parametrize $\mathcal{W}$ by a $SU(1,1)$ doublet W_i satisfying [7]

$$
W_i \overline{W}^i = 1, \tag{8.5}
$$

where $\overline{W}^i = \eta^{ij} W^*_j = (W^*_1, -W^*_2)$. Both W_1 and W_2 are chiral superfields with $U(1)$ weight 2 (the linearized superfield W_{DCBA} has $U(1)$ weight 4). Their $\theta = \overline{\theta} = 0$ components describe two real scalar fields as the coset space $SU(1,1) / U(1)$.

$N \geq 6$

The completely traceless part of (D.15) is an equation without any linear term:

$$\sum_{FED} \sum_{CBA} \text{tl} \left(\frac{1}{3} M^G_{FED} \, \overline{M}^{CBA}_G - 3 \, M^C_{FEG} \, \overline{M}^{BAG}_D - \frac{1}{3} N^{\alpha\dot\alpha}_{FEDG} \, \overline{N}^{CBAG}_{\alpha\dot\alpha} \right.$$

$$- W^\alpha_{FEG} \, \overline{M}^{GCBA}_{\alpha D} - \overline{W}^{CBG}_{\dot\alpha} \, M^{\dot\alpha A}_{GFED} - \frac{i}{3} W^\alpha_{FED} \, \mathcal{D}_{\alpha\dot\alpha} \, \overline{W}^{\dot\alpha CBA}$$

$$- \frac{i}{3} \overline{W}^{\dot\alpha CBA} \, \mathcal{D}_{\alpha\dot\alpha} \, W^\alpha_{FED} - 6 \, U^C_{\alpha\dot\alpha \, F} \, W^\alpha_{EDG} \, \overline{W}^{\dot\alpha BAG}$$

$$\left. + U^C_{\alpha\dot\alpha \, G} \, W^\alpha_{FED} \, \overline{W}^{\dot\alpha GBA} + U^G_{\alpha\dot\alpha \, F} \, W^\alpha_{GED} \, \overline{W}^{\dot\alpha CBA} \right) = 0 \,. \qquad (8.6)$$

Note that this equation is nothing but the finite version of (6.17). Consistency would require that it vanishes identically, but obviously this is not always the case. In order to derive an inconsistency, it suffices to consider the quadratic approximation

$$\sum_{FED} \sum_{CBA} \text{tl} \left(M^G_{FED} \, \overline{M}^{CBA}_G - 9 \, M^C_{FEG} \, \overline{M}^{BAG}_D - N^{\alpha\dot\alpha}_{FEDG} \, \overline{N}^{CBAG}_{\alpha\dot\alpha} \right.$$

$$- 3 \, W^\alpha_{FEG} \, \overline{M}^{GCBA}_{\alpha D} - 3 \, \overline{W}^{CBG}_{\dot\alpha} \, M^{\dot\alpha A}_{GFED}$$

$$\left. - i \, W^\alpha_{FED} \, \partial_{\alpha\dot\alpha} \, \overline{W}^{\dot\alpha CBA} - i \, \overline{W}^{\dot\alpha CBA} \, \partial_{\alpha\dot\alpha} \, W^\alpha_{FED} \right) = 0 \,. \qquad (8.7)$$

Next we separate the terms which cannot cancel each other because of their Lorentz index structure:

$$\sum_{FED} \sum_{CBA} \text{tl} \left(M^G_{FED} \, \overline{M}^{CBA}_G - 9 \, M^C_{FEG} \, \overline{M}^{BAG}_D \right) = 0 \,, \qquad (8.8)$$

$$\text{tl} \left(N^{\alpha\dot\alpha}_{FEDG} \, \overline{N}^{CBAG}_{\alpha\dot\alpha} \right) = 0 \,, \qquad (8.9)$$

$$\sum_{FED} \text{tl} \left(W^\alpha_{FEG} \, \overline{M}^{GCBA}_{\alpha D} + \frac{i}{3} W^\alpha_{FED} \, \partial_{\alpha\dot\alpha} \, \overline{W}^{\dot\alpha CBA} \right) = 0 \,. \qquad (8.10)$$

We focus on the first of these equations and decompose

$$\overline{M}^{CBA}_D = \tilde{M}^{CBA}_D + \sum_{CBA} \delta^C_D \, \overline{M}^{BA} \,, \qquad \tilde{M}^{CBA}_C = 0 \,. \qquad (8.11)$$

Equation (8.8) then splits into

$$\sum_{FED} \sum_{CBA} \text{tl} \left(\tilde{M}^G_{FED} \, \tilde{M}^{CDA}_G - 9 \, \tilde{M}^C_{FEG} \, \tilde{M}^{BAG}_D \right) = 0 \,, \qquad (8.12)$$

$$\sum_{CBA} \text{tl} \left(\tilde{M}^C_{FED} \, \overline{M}^{BA} \right) = 0 \,. \qquad (8.13)$$

The last equation implies

$$\widetilde{M}_D^{CBA} = 0 \qquad \text{or} \qquad \overline{M}^{BA} = 0 \qquad (N \geq 6) \tag{8.14}$$

in the linear approximation. These restrictions are not covariant under super-Weyl transformations. From the transformation law (5.23) one obtains

$$\text{tl}\left(\overline{W}_{\dot{\alpha}}^{CBA} D_D^{\dot{\alpha}} H\right) = 0 \qquad \text{or} \qquad \overline{W}_{\dot{\alpha}}^{BAC} D_C^{\dot{\alpha}} H = 0 \qquad (N \geq 6). \tag{8.15}$$

This implies

$$\overline{W}_{\dot{\alpha}}^{CBA} = 0 \qquad \text{or} \qquad D_{\alpha}^A H = 0 \qquad (N \geq 6). \tag{8.16}$$

The first of these equations eliminates the whole multiplet, while the second one breaks conformal supersymmetry. Hence $N \geq 6$ conformal supergravity is inconsistent.

9. Invariant Actions

Finally, we are going to construct the actions for $N \leq 4$ off-shell conformal supergravity. The actions for the linearized theories are given by [18]

$$\mathcal{S} = \int \mathrm{d}^4x \, \mathrm{d}^{2N}\theta \; W^{\alpha_1 \cdots \alpha_{4-N}} W_{\alpha_1 \cdots \alpha_{4-N}} + \text{c.c.} \qquad (N = 1, \ldots, 4), \qquad (9.1)$$

where

$$D_A^{\dot{\alpha}} W_{\alpha_1 \cdots \alpha_{4-N}} = 0. \qquad (9.2)$$

In the following we shall try to generalize these expressions to the non-linear case.

9.1 Chiral Superfields

A chiral superfield Φ_A satisfies the constraint

$$\mathcal{D}_A^{\dot{\alpha}} \Phi_B = 0. \qquad (9.3)$$

The consistency condition for this constraint is $\{\mathcal{D}_A^{\dot{\alpha}}, \mathcal{D}_B^{\dot{\beta}}\} \Phi_C = 0$, i.e.,

$$T_{ABD}^{\dot{\alpha}\dot{\beta}\delta} \mathcal{D}_\delta^D \Phi_C + R_{ABC}^{\dot{\alpha}\dot{\beta}}{}^{D} \Phi_D = 0. \qquad (9.4)$$

This condition severely restricts the possible chiral superfields. For $N = 1$, a chiral superfield $\Phi_{\alpha_1 \cdots \alpha_n}$ may have an arbitrary number of undotted spinor indices and an arbitrary U(1) weight. For $N = 2$, Φ must be a Lorentz and SU(2) scalar, but it may again have an arbitrary U(1) weight. The only exception of this is $W_{\alpha\beta}$. For $N = 3$ and $N = 4$, the only chiral superfields are W_α and W_i, respectively, whereas for $N > 4$ no chiral superfields exist (except the constants, of course).

In the cases where the condition (9.4) vanishes identically, the constraint (9.3) can be solved by an unconstrained prepotential V_A:

$$\Phi_A = \overline{\Delta} V_A. \qquad (9.5)$$

$\overline{\Delta}$ is the so-called chiral projection operator. For $N = 1$, it is given by [19]

$$\overline{\Delta} = \mathcal{D}_{\dot{\alpha}} \mathcal{D}^{\dot{\alpha}} + 4\overline{S} \qquad (9.6)$$

and for $N = 2$ one finds

$$\overline{\Delta} = \mathcal{D}_{\dot{\alpha}A} \mathcal{D}_B^{\dot{\alpha}} \left(\mathcal{D}_{\dot{\beta}}^A \mathcal{D}^{\dot{\beta}B} + 16 \, \overline{S}^{AB} \right)$$
$$- \mathcal{D}_{\dot{\alpha}A} \mathcal{D}_{\dot{\beta}}^A \left(\mathcal{D}_B^{\dot{\alpha}} \mathcal{D}^{\dot{\beta}B} - 16 \, \overline{V}^{\dot{\alpha}\dot{\beta}} \right), \qquad (9.7)$$

where $\overline{V}_{\dot{\beta}\dot{\alpha}}^{BA} = \varepsilon^{BA} \overline{V}_{\dot{\beta}\dot{\alpha}}$. There are no chiral projection operators for $N > 2$.

9.2 Superfield Actions

An arbitrary variation (3.1) of the vielbein changes its determinant (2.17) as follows:

$$\delta E = (-)^a H_A{}^A E. \tag{9.8}$$

In particular, for general coordinate transformations one obtains from (2.29) and (5.1):

$$\delta E = (-)^m \partial_M (\xi^M E)$$
$$= (-)^a (\mathcal{D}_A \xi^A + \xi^B T_{BA}{}^A) E. \tag{9.9}$$

The product of E and a scalar superfield V transforms as

$$\delta_\xi (VE) = \delta_{\xi V} E. \tag{9.10}$$

Thus the general form of an invariant superfield action is

$$\mathcal{S} = \int dz\, E V, \tag{9.11}$$

where $dz = d^4x\, d^{4N}\theta$. Another consequence of (9.9) is the formula for partial integration in superspace [20]:

$$\int dz\, E\, (-)^a (\mathcal{D}_A V^A + V^B T_{BA}{}^A) = 0. \tag{9.12}$$

In particular, this implies

$$\int dz\, E\, \mathcal{D}_\alpha^A V_A^\alpha = 0. \tag{9.13}$$

In conformal supergravity, the action (9.11) has to be invariant under super-Weyl transformations, too. That is, the variation

$$\delta E = 2(N-2) H E \tag{9.14}$$

has to be compensated by δV. Hence V must have negative dimension for $N > 2$. This is the main obstacle in constructing superfield actions. We shall discuss the various cases in more detail in the following.

$N = 1$. For $N = 1$, the basic chiral superfield $W_{\alpha\beta\gamma}$ can be written as

$$W_{\alpha\beta\gamma} = \overline{\Delta}\, V_{\alpha\beta\gamma}. \tag{9.15}$$

The non-linear generalization of the action (9.1) is then given by

$$\mathcal{S} = \int dz\, E\, V^{\alpha\beta\gamma} W_{\alpha\beta\gamma} + \text{c. c.}. \tag{9.16}$$

The prepotential $V_{\alpha\beta\gamma}$ may be eliminated by "inverting" (9.15),

$$V_{\alpha\beta\gamma} = \frac{1}{4} \overline{S}^{-1} W_{\alpha\beta\gamma} - \frac{1}{4} \mathcal{D}_{\dot\alpha} \mathcal{D}^{\dot\alpha} (\overline{S}^{-1} V_{\alpha\beta\gamma}), \tag{9.17}$$

and integrating by parts (9.13). Up to a factor, the action (9.16) becomes [19]

$$\mathcal{S} = \int dz \, E \, \overline{S}^{-1} \, W^{\alpha\beta\gamma} \, W_{\alpha\beta\gamma} + \text{c. c.} \tag{9.18}$$

and it is easy to check that this expression is super-Weyl invariant.

$N = 2$. For $N = 2$, E itself is super-Weyl invariant, so the obvious candidate for an invariant action is $\int dz \, E$. However, from (9.5, 7) and (9.13) follows

$$\int dz \, E \, \Phi = 0 \tag{9.19}$$

for any chiral superfield Φ. In particular, this implies

$$\int dz \, E = 0, \tag{9.20}$$

i. e., the volume of $N = 2$ superspace vanishes. This has already been shown in Ref. [21] within the chiral superspace approach.

In order to find an action for $N = 2$ conformal supergravity, one might now try to proceed in analogy to the case $N = 1$. However, $W_{\alpha\beta} = \overline{\Delta} \, V_{\alpha\beta}$ does not work because the right-hand side does not satisfy the condition (9.4). Another possibility would be to set $W^{\alpha\beta} \, W_{\alpha\beta} = \overline{\Delta} \, V$, but this is not invertible like (9.17) because $\overline{\Delta}$ is a total derivative. For $N > 2$, the situation is even worse since there is no chiral projection operator.

Thus there seems to be no way to construct superfield actions for $N > 1$ conformal supergravity which do not explicitly contain the prepotentials. On the other hand, the solution of the constraints in terms of unconstrained prepotentials is a very tedious procedure for higher N. Therefore we turn to another method which yields a more direct generalization of the linearized results.

9.3 Chiral Actions [1]

Chiral Coordinates. The action (9.1) for linearized conformal supergravity suggests the introduction of chiral coordinates (x, Θ) instead of $(x, \theta, \overline{\theta})$. For any chiral superfield Φ_B we define

$$\widehat{\Phi}_B(z, \Theta) = \exp\left(\Theta_A^{\alpha} \, \mathcal{D}_{\alpha}^A\right) \Phi_B(z)$$

$$= \sum_{n=0}^{2N} \frac{1}{n!} \, \Theta_{A_1}^{\alpha_1} \ldots \Theta_{A_n}^{\alpha_n} \, \mathcal{D}_{\alpha_n}^{A_n} \ldots \mathcal{D}_{\alpha_1}^{A_1} \, \Phi_B(z). \tag{9.21}$$

The old θ-variables may now be set equal to zero, but it is more convenient to keep them as long as possible.

---

[1]   This section is based on the work of Ramirez [22], which, in turn, is based on a paper by Wess and Zumino [23].

The transformation law of $\widehat{\Phi}_B$ follows from (2.32). Since Θ_A^α is a constant spinor, one finds

$$\delta\widehat{\Phi}_B = \xi^A \mathcal{D}_A \widehat{\Phi}_B - \Theta_C^\gamma \Lambda_{\gamma A}^{C\alpha} \partial_\alpha^A \widehat{\Phi}_B - \Lambda_B{}^A \widehat{\Phi}_A, \tag{9.22}$$

where $\partial_\alpha^A = \partial/\partial\Theta_A^\alpha$. Next one replaces the covariant derivatives $\mathcal{D}_A$ by partial derivatives acting on the chiral coordinates (x, Θ).

THEOREM [22]: The transformation law (9.22) can be written in the form

$$\delta\widehat{\Phi}_B = \widehat{\xi}^{\underline{m}}(z, \Theta)\, \partial_{\underline{m}} \widehat{\Phi}_B - \widehat{\Lambda}_B{}^A(z, \Theta)\, \widehat{\Phi}_A, \tag{9.23}$$

where $\partial_{\underline{m}} = (\partial_m, \partial_\alpha^A)$.

PROOF: We shall prove this theorem by induction in the power of Θ. In the first step, we show that

$$\delta\widehat{\Phi}_B = \widehat{\eta}^a \mathcal{D}_a \widehat{\Phi}_B + \widehat{\eta}_A^\alpha \partial_\alpha^A \widehat{\Phi}_B - \widehat{\lambda}_B{}^A \widehat{\Phi}_A, \tag{9.24}$$

where

$$\widehat{\eta}^a = \xi^a + \xi^{\underline{\alpha}} \widehat{\eta}_{\underline{\alpha}}{}^a,$$

$$\widehat{\eta}_A^\alpha = \xi^{\underline{\beta}} \widehat{\eta}_{\underline{\beta}A}{}^\alpha - \Theta_B^\beta \Lambda_{\beta A}^{B\alpha},$$

$$\widehat{\lambda}_B{}^A = \Lambda_B{}^A + \xi^{\underline{\gamma}} \widehat{\lambda}_{\gamma B}{}^A. \tag{9.25}$$

At $\Theta = 0$ one finds

$$\widehat{\eta}_{\underline{\alpha}}{}^a = \widehat{\eta}_{BA}^{\dot\beta\alpha} = \widehat{\lambda}_{\gamma B}{}^A = 0,$$

$$\widehat{\eta}_{\beta A}^{B\alpha} = \delta_\beta^\alpha \delta_A^B. \tag{9.26}$$

Then one takes the Θ^n-component of (9.22), commutes $\mathcal{D}_{\underline{\alpha}}$ to the right by one place, and inserts (9.24) at order Θ^{n-1}. This yields the recurrence formulas

$$\Theta\partial \widehat{\eta}_{\underline{\alpha}}{}^a + \widehat{\eta}_{\underline{\alpha}B}^\beta \widehat{\eta}_\beta^{Ba} = \Theta D \widehat{\eta}_{\underline{\alpha}}{}^a + \Theta_B^\beta T_{\beta\underline{\alpha}}^{B\ a}$$
$$+ \Theta_B^\beta (T_{\beta\underline{\alpha}}^{B\ \gamma} - \widehat{\eta}_{\underline{\alpha}}{}^b T_{b\beta}^{B\gamma}) \widehat{\eta}_{\underline{\gamma}}{}^a, \tag{9.27}$$

$$(\Theta\partial - 1) \widehat{\eta}_{\underline{\beta}A}^\alpha + \widehat{\eta}_{\underline{\beta}C}^\gamma \widehat{\eta}_{\gamma A}^{C\alpha} = \Theta D \widehat{\eta}_{\underline{\beta}A}^\alpha + \Theta_C^\gamma (T_{\gamma\underline{\beta}}^{C\ \delta} - \widehat{\eta}_{\underline{\beta}}{}^a T_{a\gamma}^{C\delta}) \widehat{\eta}_{\underline{\delta}A}^\alpha$$
$$- \Theta_C^\gamma \Theta_D^\delta (R_{\delta\underline{\beta}\gamma A}^{C\alpha} + \widehat{\eta}_{\underline{\beta}}{}^a R_{\delta a\gamma A}^{C\alpha}), \tag{9.28}$$

$$\Theta\partial \widehat{\lambda}_{\underline{\gamma}B}{}^A + \widehat{\eta}_{\underline{\gamma}D}^\delta \widehat{\lambda}_{\delta B}^D{}^A = \Theta D \widehat{\lambda}_{\underline{\gamma}B}{}^A + \Theta_D^\delta (T_{\delta\underline{\gamma}}^{D\ \varepsilon} - \widehat{\eta}_{\underline{\gamma}}{}^a T_{a\delta}^{D\varepsilon}) \widehat{\lambda}_{\underline{\varepsilon}B}{}^A$$
$$- \Theta_D^\delta (R_{\delta\underline{\gamma}B}^D{}^A + \widehat{\eta}_{\underline{\gamma}}{}^a R_{\delta aB}^D{}^A). \tag{9.29}$$

40

In the second step, we employ

$$\mathcal{D}_a = E^{-1\,m}_{a}\left(\mathcal{D}_m - E_m^{\underline{\alpha}}\,\mathcal{D}_{\underline{\alpha}}\right), \qquad E^{-1\,m}_{a}\,E_m^{b} = \delta_a^{\,b} \tag{9.30}$$

and

$$\mathcal{D}_m\widehat{\varPhi}_B = \partial_m\widehat{\varPhi}_B - \varTheta_C^{\gamma}\,\varPhi_{m\,\gamma A}^{C\alpha}\,\partial_\alpha^A\widehat{\varPhi}_B - \varPhi_{mB}^{A}\,\widehat{\varPhi}_A \tag{9.31}$$

to bring the transformation law into its final form (9.23). The parameters $\widehat{\xi}$ and $\widehat{\varLambda}$ are given by

$$\widehat{\xi}^{\,m} = \widehat{\eta}^{\,a}\,\widehat{\xi}_a^{\,m}\,,$$

$$\widehat{\xi}_A^{\,\alpha} = \widehat{\eta}_A^{\,\alpha} + \widehat{\eta}^{\,a}\,\widehat{\xi}_{aA}^{\,\alpha} - \varTheta_B^{\beta}\,\widehat{\xi}^{\,m}\,\varPhi_{m\,\beta A}^{B\alpha}\,,$$

$$\widehat{\varLambda}_B^{\,A} = \widehat{\lambda}_B^{\,A} + \widehat{\eta}^{\,c}\,\widehat{\varLambda}_{cB}^{\,A} + \widehat{\xi}^{\,m}\,\varPhi_{mB}^{A} \tag{9.32}$$

with the recurrence formulas

$$\widehat{\xi}_a^{\,m} = E^{-1\,m}_{a} - E^{-1\,n}_{a}\,E_n^{\underline{\alpha}}\,\widehat{\eta}_{\underline{\alpha}}^{\,\,b}\,\widehat{\xi}_b^{\,m}\,,$$

$$\widehat{\xi}_{aA}^{\,\alpha} = -E^{-1\,m}_{a}\,E_m^{\underline{\beta}}\left(\widehat{\eta}_{\underline{\beta}A}^{\,\alpha} + \widehat{\eta}_{\underline{\beta}}^{\,\,b}\,\widehat{\xi}_{bA}^{\,\alpha}\right),$$

$$\widehat{\varLambda}_{cB}^{\,A} = -E^{-1\,m}_{c}\,E_m^{\underline{\gamma}}\left(\widehat{\lambda}_{\underline{\gamma}B}^{\,A} + \widehat{\eta}_{\underline{\gamma}}^{\,\,d}\,\widehat{\varLambda}_{dB}^{\,A}\right). \tag{9.33}$$

Covariant Derivative. The next goal is to construct a "chiral" vielbein. To this aim we define

$$\widehat{\mathcal{D}}_{\underline{a}}\widehat{\varPhi}(z,\varTheta) = e^{\varTheta\mathcal{D}}\,\mathcal{D}_{\underline{a}}\varPhi(z), \tag{9.34}$$

where $\varPhi$ is a Lorentz and U(N) scalar. The problem of this definition is that $\mathcal{D}_{\underline{a}}\varPhi$ is not a chiral superfield:

$$\mathcal{D}_A^{\dot{\alpha}}\mathcal{D}_{\underline{b}}\varPhi = -T_{A\,\underline{b}}^{\dot{\alpha}\,\,\underline{c}}\,\mathcal{D}_{\underline{c}}\varPhi. \tag{9.35}$$

Therefore the transformation law of $\widehat{\mathcal{D}}_{\underline{a}}\widehat{\varPhi}$ is not given by (9.23). Fortunately, however, the right-hand side of (9.35) can be absorbed by a slight modification. We find

$$\delta\widehat{\mathcal{D}}_{\underline{a}}\widehat{\varPhi} = \widehat{\xi}^{\,\underline{m}}\,\partial_{\underline{m}}\,\widehat{\mathcal{D}}_{\underline{a}}\widehat{\varPhi} - \widehat{\varLambda}'^{\,\,\underline{b}}_{\underline{a}}\,\widehat{\mathcal{D}}_{\underline{b}}\widehat{\varPhi}, \tag{9.36}$$

where $\widehat{\varLambda}'^{\,\,\underline{a}}_{\underline{b}}$ is defined by the same recurrence formulas as $\widehat{\varLambda}_{\underline{b}}^{\,\underline{a}}$, but with the initial conditions

$$\widehat{\lambda}'^{\,C\,\,\underline{a}}_{\,\,\gamma\underline{b}} = 0\,,$$

$$\widehat{\lambda}'^{\,\dot{\gamma}\,\,\underline{a}}_{\,\,C\underline{b}} = T_{C\underline{b}}^{\dot{\gamma}\,\,\underline{a}} \qquad (\varTheta = 0). \tag{9.37}$$

Observe that $\widehat{\varLambda}'^{\,\,\underline{a}}_{\underline{b}}$ is no longer Lie algebra valued. The supertrace, however, remains unchanged:

$$(-)^a\,\widehat{\varLambda}'^{\,\,\underline{a}}_{\underline{a}} = (-)^a\,\widehat{\varLambda}_{\underline{a}}^{\,\underline{a}} = -2N\,\widehat{\varLambda}. \tag{9.38}$$

Chiral Vielbein. The chiral vielbein $\widehat{E}_{\underline{m}}{}^{\underline{a}}$ can now be defined as follows.

THEOREM:

$$\partial_{\underline{m}} \widehat{\Phi} = \widehat{E}_{\underline{m}}{}^{\underline{a}}(z,\Theta)\ \widehat{\mathcal{D}}_{\underline{a}} \widehat{\Phi}. \tag{9.39}$$

PROOF: From the preceding theorem one obtains

$$\mathcal{D}_{\underline{a}} \widehat{\Phi} = \widehat{e}_{\underline{a}}{}^{m}\ \partial_{\underline{m}} \widehat{\Phi}, \tag{9.40}$$

where

$$\widehat{e}_{a}{}^{m} = \widehat{\xi}_{a}{}^{m},$$

$$\widehat{e}_{a}{}^{\alpha}{}_{A} = \widehat{\xi}_{aA}^{\alpha} - \Theta^{\beta}_{B}\,\widehat{\xi}_{a}{}^{m}\,\Phi_{m\beta A}^{\ B\alpha},$$

$$\widehat{e}^{Am}_{\alpha} = \widehat{\eta}^{Aa}_{\alpha}\,\widehat{\xi}_{a}{}^{m},$$

$$\widehat{e}^{A\beta}_{\alpha B} = \widehat{\eta}^{A\beta}_{\alpha B} + \widehat{\eta}^{Aa}_{\alpha}\left(\widehat{\xi}_{aB}^{\ \beta} - \Theta^{\gamma}_{C}\,\widehat{\xi}_{a}{}^{m}\,\Phi_{m\gamma B}^{\ C\beta}\right). \tag{9.41}$$

Inverting (9.40) gives

$$\partial_{\underline{m}} \widehat{\Phi} = \widehat{e}_{\underline{m}}{}^{\underline{a}}\, \mathcal{D}_{\underline{a}} \widehat{\Phi}, \qquad \widehat{e}_{\underline{m}}{}^{\underline{a}}\,\widehat{e}_{\underline{a}}{}^{\underline{n}} = \delta_{\underline{m}}^{\underline{n}}. \tag{9.42}$$

In the next step, we prove by induction that

$$\mathcal{D}_{\underline{A}} \widehat{\Phi} = \widehat{\nabla}_{\underline{A}}{}^{\underline{b}}\, \widehat{\mathcal{D}}_{\underline{b}} \widehat{\Phi}. \tag{9.43}$$

$\widehat{\nabla}$ is given by the $\Theta = 0$ components

$$\widehat{\nabla}_{\underline{a}}{}^{\underline{b}} = \delta_{\underline{a}}^{\underline{b}}, \qquad \widehat{\nabla}_{A}^{\dot{\alpha}\,\underline{b}} = 0 \tag{9.44}$$

and the recurrence formula

$$\Theta\partial\,\widehat{\nabla}_{\underline{A}}{}^{\underline{b}} = \Theta\mathcal{D}\,\widehat{\nabla}_{\underline{A}}{}^{\underline{b}} + \Theta^{\gamma}_{C}\,T^{C}_{\gamma\underline{A}}{}^{D}\,\widehat{\nabla}_{\underline{D}}{}^{\underline{b}}$$
$$- \Theta^{\gamma}_{C}\,\Theta^{\delta}_{D}\,R_{\underline{A}\delta\gamma E}^{\ DC\varepsilon}\,\widehat{e}_{\varepsilon}^{Ef}\,\widehat{\nabla}_{\underline{f}}{}^{\underline{b}}. \tag{9.45}$$

Altogether, one finds

$$\widehat{E}_{\underline{m}}{}^{\underline{a}} = \widehat{e}_{\underline{m}}{}^{\underline{b}}\,\widehat{\nabla}_{\underline{b}}{}^{\underline{a}}, \tag{9.46}$$

which completes the proof of (9.39).

The $\Theta = 0$ component of the chiral vielbein is

$$\widehat{E}_{\underline{m}}{}^{\underline{a}} = \begin{pmatrix} E_{m}{}^{a} & E_{mA}{}^{\alpha} \\ 0 & \delta_{\beta}^{\alpha}\delta_{A}^{B} \end{pmatrix} \tag{9.47}$$

and its transformation law follows from (9.23) and (9.36):

$$\delta\widehat{E}_{\underline{m}}{}^{\underline{a}} = \widehat{\xi}^{\underline{n}}\,\partial_{\underline{n}}\,\widehat{E}_{\underline{m}}{}^{\underline{a}} + \left(\partial_{\underline{m}}\widehat{\xi}^{\underline{n}}\right)\widehat{E}_{\underline{n}}{}^{\underline{a}} + \widehat{E}_{\underline{m}}{}^{\underline{b}}\,\widehat{\Lambda}'_{\underline{b}}{}^{\underline{a}}. \tag{9.48}$$

Chiral Density. The determinant of $\widehat{E}_m{}^a$ is the chiral density

$$\mathcal{E}(z, \Theta) = \det \widehat{E}_{\underline{m}}{}^{\underline{a}}. \qquad (9.49)$$

Its $\Theta = 0$ component is

$$\mathcal{E} = \det E_m{}^a \qquad (9.50)$$

and the transformation law can be obtained from (9.48) and (9.38):

$$\delta \mathcal{E} = (-)^m \, \partial_{\underline{m}} \left(\widehat{\xi}^{\underline{m}} \, \mathcal{E} \right) - 2N \, \widehat{\Lambda} \, \mathcal{E}. \qquad (9.51)$$

The last two equations determine the Θ-expansion of $\mathcal{E}$ completely. They can be used for an explicit computation of the components of $\mathcal{E}$.

Chiral Action. In order to construct an invariant action, one has to compensate the last term of (9.51) by a chiral superfield W with U(1) weight $2N$. To be precise,

$$\delta \widehat{W} = \widehat{\xi}^{\underline{m}} \, \partial_{\underline{m}} \widehat{W} + 2N \, \widehat{\Lambda} \, \widehat{W} \qquad (9.52)$$

implies

$$\delta \left(\mathcal{E} \, \widehat{W} \right) = (-)^m \, \partial_{\underline{m}} \left(\widehat{\xi}^{\underline{m}} \, \mathcal{E} \, \widehat{W} \right). \qquad (9.53)$$

Thus the general form of a chiral action is

$$S = \int \mathrm{d}^4 x \, \mathrm{d}^{2N} \Theta \, \mathcal{E} \, \widehat{W} \Big|_{\theta = \bar{\theta} = 0} + \mathrm{c.\,c.}. \qquad (9.54)$$

Super-Weyl Invariance. At last, we analyze the super-Weyl transformations in chiral superspace. Consider a Weyl-covariant chiral superfield Φ_B with Weyl weight ω, i.e.,[1]

$$\delta \Phi_B = \omega H \Phi_B. \qquad (9.55)$$

In complete analogy to (9.23), one can prove that

$$\delta \widehat{\Phi}_B = \widehat{\zeta}^{\underline{m}} \, \partial_{\underline{m}} \widehat{\Phi}_B - \widehat{L}_B{}^A \, \widehat{\Phi}_A + \omega \widehat{H} \, \widehat{\Phi}_B, \qquad (9.56)$$

where

$$\widehat{H} = \mathrm{e}^{\Theta \mathcal{D}} H. \qquad (9.57)$$

Since the parameters $\widehat{\zeta}$ and $\widehat{L}$ can simply be added to $\widehat{\xi}$ and $\widehat{\Lambda}$, respectively, we consider only the part

$$\delta_0 \widehat{\Phi}_B = \omega \widehat{H} \, \widehat{\Phi}_B. \qquad (9.58)$$

---

[1] In general, the chirality constraint (9.3) breaks this super-Weyl invariance. This is not the case for the examples given at the end of this section.

The covariant derivative (9.34) transforms as

$$\delta_0 \widehat{\mathcal{D}}_{\underline{a}} \widehat{\Phi} = -\widehat{H}_{\underline{a}}{}^{\underline{b}} \, \widehat{\mathcal{D}}_{\underline{b}} \, \widehat{\Phi} \,, \tag{9.59}$$

where $\omega(\Phi) = 0$ and $\widehat{H}_{\underline{b}}{}^{\underline{a}} = e^{\Theta \mathcal{D}} H_{\underline{b}}{}^{\underline{a}}$. $H_{\underline{b}}{}^{\underline{a}}$ is given by (5.10–13). This implies for the chiral vielbein

$$\delta_0 \widehat{E}_{\underline{m}}{}^{\underline{a}} = \widehat{E}_{\underline{m}}{}^{\underline{b}} \, \widehat{H}_{\underline{b}}{}^{\underline{a}} \tag{9.60}$$

and for the chiral density

$$\delta_0 \mathcal{E} = (N - 4) \, \widehat{H} \, \mathcal{E} \,. \tag{9.61}$$

Hence the chiral action (9.54) is super-Weyl invariant if W is Weyl-covariant and

$$\omega(W) = 4 - N \,. \tag{9.62}$$

Conformal Actions. It is now easy to find the non-linear actions for $N \leq 4$ off-shell conformal supergravity. The proper generalization of (9.1) is

$$N = 1: \qquad S = \int \mathrm{d}^4 x \, \mathrm{d}^2 \Theta \, \mathcal{E} \, \widehat{W}^{\alpha\beta\gamma} \, \widehat{W}_{\alpha\beta\gamma} \big| + \mathrm{c.c.}, \tag{9.63}$$

$$N = 2: \qquad S = \int \mathrm{d}^4 x \, \mathrm{d}^4 \Theta \, \mathcal{E} \, \widehat{W}^{\alpha\beta} \, \widehat{W}_{\alpha\beta} \big| + \mathrm{c.c.}, \tag{9.64}$$

$$N = 3: \qquad S = \int \mathrm{d}^4 x \, \mathrm{d}^6 \Theta \, \mathcal{E} \, \widehat{W}^{\alpha} \, \widehat{W}_{\alpha} \big| + \mathrm{c.c.}, \tag{9.65}$$

$$N = 4: \qquad S = \int \mathrm{d}^4 x \, \mathrm{d}^8 \Theta \, \mathcal{E} \, \widehat{f}(W_i) \big| + \mathrm{c.c.}. \tag{9.66}$$

(To remind the reader of the notation: the hat introduces the new Θ-variables and the vertical bar removes the old θ's.) The function $f(W_i)$ in (9.66) must have U(1) weight 8. Since W_i has U(1) weight 2, this is equivalent to

$$W_i \frac{\partial f}{\partial W_i} = 4f \,. \tag{9.67}$$

The general solution of this differential equation is

$$f = (W_1 W_2)^2 \, g\!\left(\frac{W_1}{W_2}\right). \tag{9.68}$$

Thus the $N = 4$ action includes an arbitrary holomorphic function g. The global SU(1, 1) symmetry is always broken by the action, simply because $\varepsilon^{ij} W_i W_j = 0$.

Note that, in contrast to actions like (9.18), it is straightforward to compute the component form of (9.63–66) using the recurrence formulas given in this section. For obvious reasons we refrain from doing this here. On the other hand, it is easy to extract some particular informations. For instance, the linearized lagrangian is located in the

Θ^{2N}-component of $\widehat{W}\,\widehat{W}$, resp. $\widehat{f}$. In the case $N = 4$, the square of the Weyl tensor is multiplied by the real part of

$$\varepsilon_{ik}\,\varepsilon_{jl}\,\overline{W}^i\,\overline{W}^j\,\frac{\partial^2 f}{\partial W_k\,\partial W_l}\bigg|.\tag{9.69}$$

The imaginary part of (9.69) is the factor in front of the Hirzebruch invariant $\mathrm{tr}(RR)$.

It remains to be shown that the field equations for $N \leq 4$ conformal supergravity are consistent. Fortunately, this follows already from the existence of the off-shell actions. The argument goes as follows. Assume that there are some field equations from which one can derive an equation without a (or with a wrong) linear term. Then we know by off-shell supersymmetry that there must be a component field such that the variation of the action with respect to this field yields the inconsistent equation. On the other hand, we know from the linearized theory that this cannot be the case. Hence $N \leq 4$ on-shell conformal supergravity is consistent.

9.4 Summary

Let us summarize the most important results of this chapter. The actions for $N \leq 4$ conformal supergravity can be written as integrals over a chiral subspace of conventional extended superspace. The general form of a chiral action is

$$\mathcal{S} = \int \mathrm{d}^4x\,\mathrm{d}^{2N}\Theta\;\mathcal{E}\,\widehat{W}\bigg|_{\theta=\bar{\theta}=0} + \text{c.c.}.\tag{9.70}$$

$\mathcal{E}(z,\Theta)$ is the chiral density. Its Θ-expansion is uniquely determined by the lowest component

$$\mathcal{E} = \det E_m{}^a \qquad (\Theta = 0)\tag{9.71}$$

and the transformation law

$$\delta\mathcal{E} = (-)^m\,\partial_{\underline{m}}\big(\widehat{\xi}^{\underline{m}}\,\mathcal{E}\big) - 2N\,\widehat{\Lambda}\,\mathcal{E}.\tag{9.72}$$

The parameters $\widehat{\xi}$ and $\widehat{\Lambda}$ are defined by (9.32–33) in terms of $\widehat{\eta}$ and $\widehat{\lambda}$. The parameters $\widehat{\eta}$ and $\widehat{\lambda}$, in turn, are given by (9.25–29).

The superfield $\widehat{W}(z,\Theta)$ is defined by

$$\widehat{W}(z,\Theta) = \exp\big(\Theta_A^\alpha\,\mathcal{D}_\alpha^A\big)\,W(z),\tag{9.73}$$

where $W(z)$ is a chiral superfield with U(1) weight $2N$. Alternatively, $\widehat{W}$ may be defined by the $\Theta = 0$ component

$$\widehat{W} = W \qquad (\Theta = 0)\tag{9.74}$$

and the transformation law

$$\delta\widehat{W} = \widehat{\xi}^{\underline{m}}\,\partial_{\underline{m}}\widehat{W} + 2N\,\widehat{\Lambda}\,\widehat{W}.\tag{9.75}$$

Because of

$$\delta\left(\mathcal{E}\,\widehat{W}\right) = (-)^{m}\,\partial_{\underline{m}}\left(\widehat{\xi}^{\,\underline{m}}\,\mathcal{E}\,\widehat{W}\right), \tag{9.76}$$

the action (9.70) is invariant under general coordinate and structure group transformations. It is in addition invariant under super-Weyl transformations if W is Weyl-covariant with Weyl weight $4 - N$.

The actions for $N \leq 4$ conformal supergravity are then given by (9.70), where

$$W = \begin{cases} W^{\alpha\beta\gamma}\,W_{\alpha\beta\gamma} & (N = 1), \\ W^{\alpha\beta}\,W_{\alpha\beta} & (N = 2), \\ W^{\alpha}\,W_{\alpha} & (N = 3), \\ f\left(W_{i}\right) & (N = 4). \end{cases} \tag{9.77}$$

The function f satisfies the differential equation

$$W_{i}\,\frac{\partial f}{\partial W_{i}} = 4f \tag{9.78}$$

with the general solution

$$f = \left(W_{1}\,W_{2}\right)^{2}\,g\!\left(\frac{W_{1}}{W_{2}}\right). \tag{9.79}$$

Thus the conformal actions are unique for $N \leq 3$, whereas the $N = 4$ action contains an arbitrary holomorphic function g.

Part II

Poincaré Supergravity

10. Introduction and Summary

10.1 Brief Survey

In the second part of these notes we shall deal with Poincaré supergravity theories. They are extensions of the general theory of relativity and therefore of more phenomenological interest than the conformal supergravities. On the other hand, the conformal theories are a convenient starting point for the derivation of the corresponding Poincaré theories since they have more symmetries than the latter ones. In other words, $N \leq 4$ Poincaré supergravity may simply be considered as broken conformal supergravity. There are, however, also three "exceptional" Poincaré supergravities ($N = 5, 6, 8$) which are not related to any conformal theory.

On-shell Poincaré supergravity theories were constructed from $N = 1$ up to $N = 8$ (see Ref. [24] for a review). The $N \leq 4$ theories were given in Refs. [25], [26], [27], and [28, 29], respectively. Partial results for $N = 8$ were found in Ref. [30] and the full $N = 8$ theory was derived in Ref. [31] by dimensional reduction of eleven-dimensional supergravity. All the $N < 8$ Poincaré supergravities are truncations of this theory. There are no supergravity theories for $N > 8$ since they would contain fields with spin > 2, which cannot be consistently coupled to the graviton [32].

While the situation for the on-shell theories is quite clear, much less is known about off-shell Poincaré supergravity. In fact, off-shell multiplets were found only for $N = 1$ [33–38] and $N = 2$ [39, 6, 40–44]. In Ref. [45] it was argued that there are no off-shell theories without central charges for $N \geq 3$, but these no-go theorems do not exclude all possibilities (e. g., $N = 3$ supergravity coupled to three vector multiplets or $N = 4$ supergravity coupled to six vector multiplets).

Although most of the above theories were first derived from a component formalism in ordinary space-time, they can also be formulated as geometrical theories in superspace (see Ref. [12] for a review). Both methods are equivalent, but the structure of the various supergravities becomes more transparent in the geometrical approach. Moreover, this is probably the only approach which permits a classification of all consistent theories. In the following we shall start from essentially the same assumptions that we have already used in the conformal case and we shall show that they reproduce exactly the known spectrum of Poincaré supergravity theories.

10.2 Assumptions

Our classification of consistent Poincaré supergravity theories is based on the same assumptions as in the conformal case (Section 1.2). In order to exclude $N = 3$ and $N = 4$ off-shell Poincaré supergravities, we need in addition the following assumption:

(C) Two more covariant (super)fields with dimension zero and high spins vanish (Eq. (13.20)).

For *on-shell* Poincaré supergravity, assumptions (A)–(C) are rather weak and can probably not be circumvented. *Off-shell*, however, there are two known exceptions:

(i) central charges [3, 46],

(ii) harmonic superspace [47].

Both possibilities require additional bosonic coordinates in superspace and thus circumvent assumption (A). Unfortunately, they also have certain drawbacks. Central charges make it impossible to construct unconstrained prepotentials. Hence theories with off-shell central charges are not much better than on-shell theories. Harmonic superspace does allow unconstrained prepotentials (by introducing an infinite number of auxiliary fields), but off-shell multiplets have been found only for $N = 2$ (supergravity and matter) and $N = 3$ (matter).

10.3 Results

The consistent Poincaré supergravity theories satisfying assumptions (A)–(C) are listed in Tables 4 and 5.

N	1	2	3	4	5	6	7	8	≥ 9
off-shell	(see Table 5)	—	—	—	—	—	—	—	—
on-shell	2	4	8	16	32	64	—	128	—

Table 4. Consistent Poincaré supergravity theories

Again, the numbers are the bosonic (fermionic) degrees of freedom of the minimal multiplets. "Consistent" is defined as in Section 1.3. On the one hand, this means that we did not care about on-shell actions, although they are known for all theories. On the other hand, it means that $N > 8$ supergravity is already algebraically inconsistent, independently of the existence of an action.

In Table 5 we have listed the minimal $N = 1$ and $N = 2$ off-shell Poincaré multiplets, their off-shell and on-shell degrees of freedom, and their off-shell local symmetries (the corresponding gauge fields are auxiliary fields). The relations of these multiplets to the conformal multiplets and among themselves are shown in Figure 1 on p. 83 and Figure 2 on p. 86.

The two $N = 1$ multiplets are the minimal multiplet [34] and the new minimal multiplet [36]. For $N = 2$, the results are rather curious. The minimal $40 + 40$ multiplet [39, 6] is reducible to the minimal $32 + 32$ multiplet [42] and the new minimal $40 + 40$

48

N	1		2			
off-shell	12	12	40	40	32	32
on-shell	2	2	4	4	8	8
local symmetry	—	U(1)	—	SO(2)	—	SO(2)
multiplet	(16.8)	(16.15)	(15.35)	(15.63)	(15.43)	(15.74)

Table 5. $N = 1, 2$ off-shell Poincaré supergravities

multiplet [41] is reducible to the new minimal $32 + 32$ multiplet [43]. On-shell, however, the $40 + 40$ multiplets are irreducible, whereas the $32 + 32$ multiplets include a physical scalar, resp. vector multiplet.

Probably all non-minimal off-shell multiplets are reducible to one of the above minimal ones. We have no proof of this,[1] but it is true for all the known theories. For instance, the $N = 1$ non-minimal $20 + 20$ multiplet [33] is reducible to a $16 + 16$ multiplet [38] and the $16 + 16$ multiplet was shown to be reducible to the minimal multiplet in Ref. [48]. The reducibility of the other non-minimal $N = 1$ multiplets [35, 37] was demonstrated in Refs. [49, 41, 50]. (The remaining $N = 2$ multiplets [40, 44] have an off-shell central charge and an infinite number of auxiliary fields, respectively.)

---

[1]   The question is not of the utmost importance either (see Chapter 17).

11. Upper Bounds

In Chapter 8 we derived an equation without any linear term. This equation limits not only the consistent conformal supergravity theories but yields also an upper bound for the Poincaré theories.

Consider Eq. (8.9) which is non-trivial only for $N \geq 7$. Because of the global SU(N) invariance of the particle spectrum [51], this equation implies on-shell

$$\overline{N}_{\alpha\dot\alpha}^{DCBA} = 0 \qquad\qquad (N = 7,\ N > 8), \qquad (11.1)$$

$$N_{\alpha\dot\alpha\,ABCD} = \frac{1}{4!}\,\varepsilon_{ABCDEFGH}\,\overline{N}_{\alpha\dot\alpha}^{EFGH} \qquad (N = 8) \qquad (11.2)$$

in the linear approximation (see Appendix C for the case $N = 8$). An eventual minus sign in (11.2) may be absorbed by the ε-tensor.

We now focus on the first of these equations and assume $N = 7$ or $N > 8$. From (7.2) we obtain

$$\overline{M}_{\alpha\,E}^{DCBA} = 0, \qquad (11.3)$$

$$\sum_{\dot\beta\dot\alpha} \partial_{\alpha\dot\beta}\,\overline{W}_{\dot\alpha}^{CBA} = 0. \qquad (11.4)$$

Equation (8.10) then becomes

$$\mathrm{tl}\left(W_{FED}^{\alpha}\,\partial_{\alpha\dot\alpha}\,\overline{W}^{\dot\alpha\,CBA} \right) = 0, \qquad (11.5)$$

which implies

$$\partial_{\alpha\dot\alpha}\,\overline{W}^{\dot\alpha\,CBA} = 0. \qquad (11.6)$$

Altogether, we find

$$\partial_{\alpha\dot\alpha}\,\overline{W}_{\dot\beta}^{CBA} = 0. \qquad (11.7)$$

This equation eliminates the x-dependence of the whole multiplet. Hence $N = 7$ and $N > 8$ Poincaré supergravities are inconsistent.

For off-shell theories the bound is even lower since the conformal field equations for $N > 4$ (Chapter 7) exclude off-shell Poincaré multiplets, too. The immediate conclusion is that $N > 4$ Poincaré supergravity has no off-shell formulation in conventional extended superspace.

12. On-Shell Poincaré Supergravity

The particle spectrum of $N \leq 8$ on-shell Poincaré supergravity is well known from the representation theory of the supersymmetry algebra [51, 52]. The number of states with helicity $2 - \frac{k}{2}$ ($k = 0, \ldots, N$) is

$$N \leq 7 : \qquad \binom{N}{k} + \binom{N}{8-k},$$

$$N = 8 : \qquad \binom{8}{k}. \tag{12.1}$$

It is easy to see that the $N = 7$ and $N = 8$ multiplets are identical, i.e., there is no supergravity theory with seven gravitinos. This agrees with the conclusions of the preceding chapter. Since the $N \leq 6$ supergravities are simply truncations of the $N = 8$ theory [31], we shall in the following only consider the case $N = 8$.

12.1 Linearized $N = 8$ Supergravity

In superspace, linearized $N = 8$ supergravity is defined by two constraints. The first one breaks the extra conformal symmetries. It can simply be stated as the vanishing of the U(8) curvature,

$$R_{DC}{}^{B}{}_{A} = 0, \tag{12.2}$$

which is equivalent to

$$V_{\beta\alpha}^{BA} = S^{BA} = U_{\alpha\dot{\alpha}A}^{B} = \overline{M}_{D}^{CBA} = 0. \tag{12.3}$$

The second constraint is the consistency condition (11.2), which breaks the global U(1) symmetry.

The consequences of these constraints can easily be determined from the solution of the Bianchi and Ricci identities. At dimension 1 we find

$$D_{\dot{\beta}D} \overline{W}_{\dot{\alpha}}^{CBA} = \sum_{CBA} \delta_{D}^{C} \overline{W}_{\dot{\beta}\dot{\alpha}}^{BA}, \tag{12.4}$$

$$D_{\alpha}^{D} \overline{W}_{\dot{\alpha}}^{CBA} = i \overline{N}_{\alpha\dot{\alpha}}^{DCBA}. \tag{12.5}$$

At dimension $\frac{3}{2}$ one obtains

$$D_{\dot{\gamma}C} \overline{W}_{\dot{\beta}\dot{\alpha}}^{BA} = -2 \sum_{BA} \delta_{C}^{B} \overline{W}_{\dot{\gamma}\dot{\beta}\dot{\alpha}}^{A}, \tag{12.6}$$

$$D_{\alpha}^{C} \overline{W}_{\dot{\beta}\dot{\alpha}}^{DA} = -\frac{i}{2} \sum_{\dot{\beta}\dot{\alpha}} \partial_{\alpha\dot{\beta}} W_{\dot{\alpha}}^{CBA}, \tag{12.7}$$

$$D_{\dot{\beta}E} \overline{N}_{\alpha\dot{\alpha}}^{DCBA} = -\frac{1}{6} \sum_{\dot{\beta}\dot{\alpha}} \sum_{DCBA} \delta_{E}^{D} \partial_{\alpha\dot{\beta}} \overline{W}_{\dot{\alpha}}^{CBA}, \tag{12.8}$$

and the field equations

$$\overline{\Psi}^A_{\beta\alpha\dot\alpha} = \overline{\Psi}^A_{\dot\alpha} = 0, \tag{12.9}$$

$$\partial_{\alpha\dot\alpha} \overline{W}^{\dot\alpha CBA} = 0. \tag{12.10}$$

At dimension 2, this implies

$$D_{\dot\delta B} \overline{W}^A_{\dot\gamma\dot\beta\dot\alpha} = -\delta^A_B \overline{W}_{\dot\delta\dot\gamma\dot\beta\dot\alpha}, \tag{12.11}$$

$$D^B_\alpha \overline{W}^A_{\dot\gamma\dot\beta\dot\alpha} = \frac{\mathrm{i}}{3!} \sum_{\dot\gamma\dot\beta\dot\alpha} \partial_{\alpha\dot\gamma} \overline{W}^{BA}_{\dot\beta\dot\alpha}, \tag{12.12}$$

the Bianchi identity

$$\sum_{\dot\beta\dot\alpha} \partial^\alpha_{\ \dot\beta} \overline{N}^{DCBA}_{\alpha\dot\alpha} = 0, \tag{12.13}$$

and the field equations

$$P_{\beta\alpha\dot\beta\dot\alpha} = R = 0, \tag{12.14}$$

$$\partial_\alpha^{\ \dot\beta} \overline{W}^{BA}_{\dot\beta\dot\alpha} = 0, \tag{12.15}$$

$$\partial^{\alpha\dot\alpha} \overline{N}^{DCBA}_{\alpha\dot\alpha} = 0. \tag{12.16}$$

Finally, at dimensions $\frac{5}{2}$ and 3 one finds

$$D_{\dot\epsilon A} \overline{W}_{\dot\delta\dot\gamma\dot\beta\dot\alpha} = 0, \tag{12.17}$$

$$D^A_\alpha \overline{W}_{\dot\delta\dot\gamma\dot\beta\dot\alpha} = \frac{\mathrm{i}}{12} \sum_{\dot\delta\dot\gamma\dot\beta\dot\alpha} \partial_{\alpha\dot\delta} \overline{W}^A_{\dot\gamma\dot\beta\dot\alpha}, \tag{12.18}$$

and the Bianchi identities

$$\partial_\alpha^{\ \dot\gamma} \overline{W}^A_{\dot\gamma\dot\beta\dot\alpha} = 0, \tag{12.19}$$

$$\partial_\alpha^{\ \dot\delta} \overline{W}_{\dot\delta\dot\gamma\dot\beta\dot\alpha} = 0. \tag{12.20}$$

Equation (12.13) shows that the $\theta = \overline\theta = 0$ component of $\overline{N}^{DCBA}_{\alpha\dot\alpha}$ is not an independent component field but the field strength of $\binom{8}{4} = 70$ scalar fields. The solution of this Bianchi identity is given by

$$\overline{N}^{DCBA}_{\alpha\dot\alpha} = \partial_{\alpha\dot\alpha} \overline{W}^{[DCBA]}, \tag{12.21}$$

where

$$W_{ABCD} = \frac{1}{4!} \varepsilon_{ABCDEFGH} \overline{W}^{EFGH} \tag{12.22}$$

and

$$D_{\dot{\alpha}E}\,\overline{W}^{DCBA} = -\frac{1}{3}\sum_{DCBA}\delta^D_E\,\overline{W}^{CBA}_{\dot{\alpha}}\,. \tag{12.23}$$

The above superfield equations determine completely the transformation laws and field equations of linearized $N = 8$ supergravity. The basic superfields are once more listed in Table 6.

spin	2	$\frac{3}{2}$	1	$\frac{1}{2}$	0
superfield	$\overline{W}_{\dot{\delta}\dot{\gamma}\dot{\beta}\dot{\alpha}}$	$\overline{W}^A_{\dot{\gamma}\dot{\beta}\dot{\alpha}}$	$\overline{W}^{BA}_{\dot{\beta}\dot{\alpha}}$	$\overline{W}^{CBA}_{\dot{\alpha}}$	$\overline{W}^{DCBA}$

Table 6. Basic superfields of linearized $N = 8$ supergravity

$\overline{W}_{\dot{\delta}\dot{\gamma}\dot{\beta}\dot{\alpha}}\big|_{\theta=\bar{\theta}=0}$ is the Weyl tensor of the graviton, $\overline{W}^A_{\dot{\gamma}\dot{\beta}\dot{\alpha}}\big|$ is the Weyl tensor of the 8 gravitinos, $\overline{W}^{BA}_{\dot{\beta}\dot{\alpha}}\big|$ is the field strength of the 28 vector fields, $\overline{W}^{CBA}_{\dot{\alpha}}\big|$ are the 56 Weyl spinors, and $\overline{W}^{DCBA}\big|$ are the 70 scalar fields. In the non-linear theory, the scalars will be described by a more complicated geometrical structure.

The basic superfields of linearized $N \le 6$ supergravity can also be read off from the above table. The additional $\binom{N}{6}$ vector fields are described by

$$\overline{W}_{\dot{\beta}\dot{\alpha}\,ABCDEF} = \frac{1}{2}\,\varepsilon_{ABCDEFGH}\,\overline{W}^{GH}_{\dot{\beta}\dot{\alpha}} \tag{12.24}$$

and the extra $\binom{N}{5}$ Weyl spinors are given by

$$\overline{W}_{\dot{\alpha}\,ABCDE} = \frac{1}{3!}\,\varepsilon_{ABCDEFGH}\,\overline{W}^{FGH}_{\dot{\alpha}}\,. \tag{12.25}$$

Furthermore, W_{ABCD} is a complex superfield for $N \le 6$ and describes $2\binom{N}{4}$ scalar fields.

12.2 Non-Linear $N = 8$ Supergravity

The structure of non-linear $N = 8$ supergravity is uniquely fixed by the linearized theory and by the requirement of non-linear consistency. In order to find the generalization of the constraints (12.3) and (11.2), we first write down all possible non-linear modifications:

$$V^{BA}_{\beta\alpha} = c_1\,W^{BACDE}_{\beta}\,W_{\alpha\,CDE}\,, \tag{12.26}$$

$$S^{BA} = 0\,, \tag{12.27}$$

$$U^B_{\alpha\dot{\alpha}\,A} = c_2\,W_{\alpha\,ACD}\,\overline{W}^{BCD}_{\dot{\alpha}} + c_3\,\delta^B_A\,W_{\alpha\,CDE}\,\overline{W}^{CDE}_{\dot{\alpha}}\,, \tag{12.28}$$

$$\overline{M}^{CBA}_D = c_4\,W^\alpha_{DEF}\,W^{CBAEF}_\alpha\,, \tag{12.29}$$

$$N_{\alpha\dot{\alpha}\,ABCD} = \frac{1}{4!}\,\varepsilon_{ABCDEFGH}\,\overline{N}^{EFGH}_{\alpha\dot{\alpha}}\,. \tag{12.30}$$

53

In the next step we determine the constants $c_1, \ldots, c_4$. Since the last constraint breaks the U(1) invariance, we must have

$$\Phi_{\underline{\alpha}} = 0 \, . \tag{12.31}$$

This implies for the U(1) curvature

$$R^{DC}_{\delta\,\gamma} = 0 \, ,$$

$$R^{D}_{\delta\,\dot\gamma C} = 2\mathrm{i}\, \delta^{D}_{C}\, \Phi_{\delta\dot\gamma} \, . \tag{12.32}$$

Comparing this with the solution of the Bianchi identities, one finds

$$c_2 = \frac{1}{4} \, , \tag{12.33}$$

$$\Phi_{\alpha\dot\alpha} = \frac{\mathrm{i}}{4}\, c_3\, W_{\alpha\,ABC}\, \overline{W}^{ABC}_{\dot\alpha} \, . \tag{12.34}$$

Similarly, one obtains from (4.20)

$$c_1 = \frac{1}{24} \, , \qquad c_3 = -\frac{1}{24} \, , \tag{12.35}$$

and from (D.4)

$$c_4 = -\frac{1}{4} \, . \tag{12.36}$$

Although the U(1) symmetry is broken, we find it convenient to keep the full $SL(2, \mathbf{C}) \times U(8)$ covariant derivatives. In order to keep track of the U(1) weights, we define

$$\mathcal{D}_{\alpha\dot\alpha}\, \varepsilon_{ABCDEFGH} = 8\, \varepsilon_{ABCDEFGH}\, \Phi_{\alpha\dot\alpha} \, , \tag{12.37}$$

where $\Phi_{\alpha\dot\alpha}$ is given by (12.34–35). It is much simpler to observe this rule than to split off the U(1) connection in all equations.

In Appendix F we have listed the consequences of the constraints (12.26–30), i. e., the non-linear versions of (12.4–20). By evaluating all possible (anti)commutators we have verified that the algebra closes on these equations. In particular, the "inconsistent" equation (8.6) is identically satisfied.[1] Hence $N = 8$ on-shell Poincaré supergravity is algebraically consistent. The same conclusion holds for the $N \le 6$ supergravities since they are truncations of the $N = 8$ theory.

One question remains: where are the scalars in the non-linear theory? The answer was first given in Ref. [31] and an elegant superspace formulation was found in Ref. [53]. The non-linear equations can be arranged in such a way that the linearized constraint "SU(8) curvature = 0" is replaced by "$E_{7(7)}$ curvature = 0". ($E_{7(7)}$ is a non-compact real form of E_7 and SU(8) is its maximal compact subgroup.) In complete analogy to

---

[1] Using this equation, one can check that the constraints (12.26–30) are not only sufficient, but also necessary for consistency.

$N = 4$ conformal supergravity, this constraint can be solved by a superfield $\mathcal{W} \in \mathrm{E}_{7(7)}$ and the 70 scalar fields are then given by the coset space $\mathrm{E}_{7(7)}/\mathrm{SU}(8)$. (Note that $\dim \mathrm{E}_7 - \dim \mathrm{SU}(8) = 133 - 63 = 70$.) The analogous results for $N = 4$, 5, and 6 are listed in Table 7.

N	local symmetry	global symmetry	scalar manifold
8	$\mathrm{SU}(8)$	$\mathrm{E}_{7(7)}$	$\mathrm{E}_{7(7)}/\mathrm{SU}(8)$
6	$\mathrm{U}(6)$	$\mathrm{SO}^*(12)$	$\mathrm{SO}^*(12)/\mathrm{U}(6)$
5	$\mathrm{U}(5)$	$\mathrm{SU}(5,1)$	$\mathrm{SU}(5,1)/\mathrm{U}(5)$
4	$\mathrm{U}(4)$	$\mathrm{SU}(4) \times \mathrm{SU}(1,1)$	$\mathrm{SU}(1,1)/\mathrm{U}(1)$

Table 7. Symmetries of $N \geq 4$ on-shell Poincaré supergravity

Finally, we add some remarks concerning the actions for on-shell Poincaré supergravity. The problem is that an action must be a function of unconstrained fields, whereas the supergravity algebra closes only on the physical states. Therefore the on-shell actions cannot be supersymmetric in a strict sense. (Strict supersymmetry would constrain them to vanish.) It is not excluded that an enlarged algebra (e. g., with central charges or BRS generators) could solve this problem, but such extensions are outside the scope of these notes.

As long as the actions are not strictly supersymmetric, they can, of course, not be written as integrals over superspace. The simplest way to obtain their component form then seems to be the "integration" of the field equations. We have not done this here since the calculation is lengthy and the results are well known.

13. One-Form Gauge Potentials

Before we turn to $N \leq 4$ off-shell Poincaré supergravity, we have to discuss the off-shell multiplets that may appear as subsets of the supergravity multiplets. We start with the vector multiplets which describe the $\binom{N}{2}$ vector fields of $N \leq 4$ Poincaré supergravity.

13.1 Superspace Geometry

Consider an arbitrary Lie group $\mathcal{G}$ with antihermitian parameters Λ,

$$\Lambda = -\Lambda^{\dagger}. \tag{13.1}$$

An arbitrary p-form Ω in the fundamental representation of $\mathcal{G}$ transforms as

$$\delta\Omega = \Omega\Lambda. \tag{13.2}$$

As in Section 2.1, we define a covariant exterior derivative

$$\mathcal{D}\Omega = d\Omega + \Omega A. \tag{13.3}$$

The gauge potential

$$A = dz^{\mathcal{M}} A_{\mathcal{M}} = E^{\mathcal{A}} A_{\mathcal{A}} \tag{13.4}$$

is a Lie algebra valued (i.e. antihermitian) 1-form with the transformation law

$$\delta A = -\mathcal{D}\Lambda = -d\Lambda - [\Lambda, A]. \tag{13.5}$$

The field strength is defined by

$$\mathcal{D}\mathcal{D}\Omega = \Omega F, \tag{13.6}$$

$$F = dA + AA. \tag{13.7}$$

It is a Lie algebra valued 2-form,

$$F = \tfrac{1}{2} dz^{\mathcal{M}} dz^{\mathcal{N}} F_{\mathcal{N}\mathcal{M}} = \tfrac{1}{2} E^{\mathcal{A}} E^{\mathcal{B}} F_{\mathcal{B}\mathcal{A}}, \tag{13.8}$$

and satisfies the Bianchi identity

$$\mathcal{D}F = 0. \tag{13.9}$$

Next we rewrite the equations that we shall need later on, in a more explicit form. The covariant derivative (13.3) of a Lie algebra valued 0-form $V_{\mathcal{A}}$ is

$$\mathcal{D}_{\mathcal{M}} V_{\mathcal{A}} = \partial_{\mathcal{M}} V_{\mathcal{A}} - \Phi_{\mathcal{M}\mathcal{A}}{}^{\mathcal{B}} V_{\mathcal{B}} - [A_{\mathcal{M}}, V_{\mathcal{A}}] \tag{13.10}$$

and the (anti)commutator of two covariant derivatives (13.6) reads

$$[\mathcal{D}_{\mathcal{A}}, \mathcal{D}_{\mathcal{B}}\} V_{\mathcal{C}} = -T_{\mathcal{A}\mathcal{B}}{}^{\mathcal{D}} \mathcal{D}_{\mathcal{D}} V_{\mathcal{C}} - R_{\mathcal{A}\mathcal{B}\mathcal{C}}{}^{\mathcal{D}} V_{\mathcal{D}} - [F_{\mathcal{A}\mathcal{B}}, V_{\mathcal{C}}\}. \tag{13.11}$$

The structure equation (13.7) gives

$$F_{\mathcal{N}\mathcal{M}} = \sum_{\mathcal{N}\mathcal{M}} \left(\partial_{\mathcal{N}} A_{\mathcal{M}} - A_{\mathcal{N}} A_{\mathcal{M}} \right), \tag{13.12}$$

$$F_{\mathcal{B}\mathcal{A}} = \sum_{\mathcal{B}\mathcal{A}} \left(\mathcal{D}_{\mathcal{B}} A_{\mathcal{A}} + A_{\mathcal{B}} A_{\mathcal{A}} \right) + T_{\mathcal{B}\mathcal{A}}{}^{\mathcal{C}} A_{\mathcal{C}} \tag{13.13}$$

and the Bianchi identity (13.9) becomes

$$\sum_{\mathcal{C}\mathcal{B}\mathcal{A}} \left(\mathcal{D}_{\mathcal{C}} F_{\mathcal{B}\mathcal{A}} + T_{\mathcal{C}\mathcal{B}}{}^{\mathcal{D}} F_{\mathcal{D}\mathcal{A}} \right) = 0. \tag{13.14}$$

Finally, the transformation law (13.5) reads explicitly

$$\delta A_{\mathcal{M}} = \xi^{\mathcal{N}} \partial_{\mathcal{N}} A_{\mathcal{M}} + \left(\partial_{\mathcal{M}} \xi^{\mathcal{N}} \right) A_{\mathcal{N}} - \mathcal{D}_{\mathcal{M}} \Lambda. \tag{13.15}$$

With the new parameter

$$\lambda = \Lambda - \xi^{\mathcal{M}} A_{\mathcal{M}} \tag{13.16}$$

it can be written in the more covariant form

$$\delta A_{\mathcal{M}} = \xi^{\mathcal{A}} F_{\mathcal{A}\mathcal{M}} - \mathcal{D}_{\mathcal{M}} \lambda. \tag{13.17}$$

13.2 Constraints

As in supergravity, one has to impose covariant constraints on the Yang-Mills field strength in order to be able to construct invariant actions. For this purpose, we decompose the lowest-dimensional components $F_{\beta\alpha}^{BA}$ and $F_{\beta A}^{B\dot{\alpha}}$ into

$$F_{\beta\,\alpha}^{BA} = F_{(\beta\alpha)}^{(BA)} + \varepsilon_{\beta\alpha} F^{[BA]},$$

$$F_{\beta\,\dot{\alpha}A}^{B} = \widetilde{F}_{\beta\,\dot{\alpha}A}^{B} + \delta_{A}^{B} F_{\beta\dot{\alpha}}, \tag{13.18}$$

where $\widetilde{F}$ is traceless in BA. The structure equation (13.13) shows that $F_{\beta\dot{\alpha}}$ can be absorbed by a redefinition of A_{a}. Therefore we impose the conventional constraint

$$F_{\beta\dot{\alpha}} = 0. \tag{13.19}$$

Furthermore, we require

$$F_{(\beta\alpha)}^{(BA)} = 0,$$

$$\widetilde{F}_{\beta\,\dot{\alpha}A}^{B} = 0 \tag{13.20}$$

in order to exclude fields with spin > 1 from the multiplet. Note that these equations correspond exactly to the constraints (3.9) for supergravity.

To summarize, the Yang-Mills constraints are

$$F_{\beta\,\alpha}^{\,BA} = \varepsilon_{\beta\alpha}\, F^{[BA]}\,,$$

$$F_{\beta\,\dot\alpha A}^{\,B} = 0\,,$$

$$F_{\dot\beta B\,\dot\alpha A} = \varepsilon_{\dot\beta\dot\alpha}\, \overline{F}_{[BA]}\,. \tag{13.21}$$

They define the $N = 1, 2$ off-shell and $N = 3$ on-shell vector multiplets. For $N = 4$, one has to impose an additional constraint because of consistency problems, and for $N > 4$ the constraints (13.21) are inconsistent. To obtain these results, we shall now analyze the Bianchi and Ricci identities.

13.3 Bianchi Identities

We start by solving the Bianchi identities (13.14) subject to the constraints (13.21).

dim $\frac{3}{2}$: $\left(\underline{\gamma\beta\,\alpha}\right)$

At dimension $\frac{3}{2}$ we obtain from $\left(^{CBA}_{\gamma\beta\alpha}\right)$

$$\mathcal{D}_\alpha^C\, F^{BA} = F_\alpha^{[CBA]} \tag{13.22}$$

and from $\left(^{CB\,\dot\alpha}_{\gamma\beta\,A}\right)$

$$F_{\beta\,\alpha\dot\alpha}^{\,B} = \mathrm{i}\,\varepsilon_{\beta\alpha}\,\overline{F}_{\dot\alpha}^{\,B}\,, \tag{13.23}$$

$$\mathcal{D}_{\dot\alpha C}\, F^{BA} = 2\sum_{BA} \delta_C^B\, \overline{F}_{\dot\alpha}^{\,A} + \overline{F}_{CD}\, \overline{W}_{\dot\alpha}^{DBA}\,. \tag{13.24}$$

dim 2: $\left(\underline{\gamma\beta\,\alpha}\right)$

At dimension 2, $\left(^{CB}_{\gamma\beta\,\alpha}\right)$ gives

$$\mathcal{D}_\alpha^B\, \overline{F}_{\dot\alpha}^{\,A} = -\mathrm{i}\, \mathcal{D}_{\alpha\dot\alpha}\, F^{BA} + F^{AC}\, U_{\alpha\dot\alpha C}^{\,B} - F_{\alpha C}\, \overline{W}_{\dot\alpha}^{CBA}\,. \tag{13.25}$$

Using the decomposition

$$F_{\beta\dot\beta\,\alpha\dot\alpha} = \varepsilon_{\dot\beta\dot\alpha}\, F_{(\beta\alpha)} - \varepsilon_{\beta\alpha}\, \overline{F}_{(\dot\beta\dot\alpha)}\,, \tag{13.26}$$

58

one obtains from $\left(\begin{smallmatrix} C\dot\beta \\ \gamma B a \end{smallmatrix}\right)$

$$\mathcal{D}_{\dot\beta B}\,\overline{F}_{\dot\alpha}^{\ A} = 2\,\delta_B^A\,\overline{F}_{\dot\beta\dot\alpha} - F^{AC}\,\overline{V}_{\dot\beta\dot\alpha\,CB} + \overline{F}_{BC}\,\overline{W}_{\dot\beta\dot\alpha}^{\ CA}$$

$$+\,\varepsilon_{\dot\beta\dot\alpha}\,\big(\,F_B^A - \tfrac{1}{2}\,F^{AC}\,\overline{S}_{CB} + \tfrac{1}{2}\,\overline{F}_{BC}\,S^{CA}\,\big)\,, \qquad (13.27)$$

where

$$F_B^A = \overline{F}_B^A\,. \qquad (13.28)$$

dim $\tfrac{5}{2}$: $\left(\begin{smallmatrix} \\ \gamma b a \end{smallmatrix}\right)$

The only dim-$\tfrac{5}{2}$ identity reads

$$\mathcal{D}_\gamma^C F_{ba} = \sum_{ba}\big(\mathcal{D}_b\,F_{\gamma a}^C + T_{b\gamma}^{\ \ C\dot\delta}\,F_{\underline{\dot\delta}a}\big) - T_{ba\,D}^{\ \ \ \delta}\,F_{\delta\gamma}^{DC}\,, \qquad (13.29)$$

which is equivalent to

$$\mathcal{D}_\gamma^A F_{\beta\alpha} = -F^{AB}\,W_{\gamma\beta\alpha\,B} - \sum_{\beta\alpha}\Big[\,V_{\gamma\beta}^{AB}\,F_{\alpha B}$$

$$+\,\varepsilon_{\gamma\beta}\,\Big(\tfrac{\mathrm{i}}{2}\,\mathcal{D}_{\alpha\dot\alpha}\,\overline{F}^{\dot\alpha A} - S^{AB}\,F_{\alpha B} - F^{AB}\,\Psi_{\alpha B}\Big)\Big]\,, \qquad (13.30)$$

$$\mathcal{D}_\alpha^A \overline{F}_{\dot\beta\dot\alpha} = F^{AB}\,\Psi_{\dot\beta\dot\alpha\,\alpha B} + \overline{W}_{\dot\beta\dot\alpha}^{AB}\,F_{\alpha B}$$

$$-\sum_{\dot\beta\dot\alpha}\Big(\tfrac{\mathrm{i}}{2}\,\mathcal{D}_{\alpha\dot\beta}\,\overline{F}_{\dot\alpha}^A + U_{\alpha\dot\beta\,B}^A\,\overline{F}_{\dot\alpha}^B\Big)\,. \qquad (13.31)$$

dim 3: $\left(\begin{smallmatrix} \\ c b a \end{smallmatrix}\right)$

At dimension 3 we are left with the identity

$$\sum_{cba}\big(\mathcal{D}_c\,F_{ba} + T_{cb}^{\ \ \dot\delta}\,F_{\underline{\dot\delta}a}\big) = 0\,, \qquad (13.32)$$

which implies

$$\mathcal{D}_{\dot\alpha}^\beta\,F_{\beta\alpha} + \mathcal{D}_\alpha^{\ \dot\beta}\,\overline{F}_{\dot\beta\dot\alpha} - \mathrm{i}\,\big(F_A^\beta\,\overline{\Psi}_{\beta\alpha\dot\alpha}^A + \overline{F}^{\dot\beta A}\,\Psi_{\dot\beta\dot\alpha\,\alpha A}\big)$$

$$-\,3\mathrm{i}\,\big(F_{\alpha A}\,\overline{\Psi}_{\dot\alpha}^A + \overline{F}_{\dot\alpha}^A\,\Psi_{\alpha A}\big) = 0\,. \qquad (13.33)$$

13.4 Ricci Identities

In order to determine the vector multiplets and their transformation laws, we now solve the Ricci identities (13.11) for the various superfields introduced in the preceding section.

dim 2

The dim-2 identity $\{\mathcal{D}_\beta^D, \mathcal{D}_\alpha^C\} F^{BA}$ gives

$$
\mathcal{D}_\beta^D F_\alpha^{CBA} = F_{(\beta\alpha)}^{[DCBA]} - \frac{1}{2} \sum_{CBA} \left(F^{DC} V_{\beta\alpha}^{BA} + F^{CB} V_{\beta\alpha}^{AD} \right)
$$

$$
- \frac{1}{3!} \varepsilon_{\beta\alpha} \sum_{CBA} \left(6 F^{CB} S^{AD} + \frac{1}{2} F^{DE} \overline{M}_E^{CBA} \right.
$$

$$
+ \frac{3}{2} F^{CE} \overline{M}_E^{BAD} + \overline{F}_{\dot\alpha}^D \overline{W}^{\dot\alpha CBA} + 3 \overline{F}_{\dot\alpha}^C \overline{W}^{\dot\alpha BAD}
$$

$$
\left. + \frac{3}{2} \overline{F}_{EF} \overline{W}_{\dot\alpha}^{EDC} \overline{W}^{\dot\alpha FBA} + \frac{3}{2} [F^{DC}, F^{BA}] \right) \tag{13.34}
$$

and $\{\mathcal{D}_\alpha^D, \mathcal{D}_{\dot\alpha C}\} F^{BA}$ implies

$$
\mathcal{D}_{\dot\alpha D} F_\alpha^{CBA} = \frac{1}{3!} \sum_{CBA} \left(- 6\mathrm{i}\, \delta_D^C \mathcal{D}_{\alpha\dot\alpha} F^{BA} + 6 F^{CB} U_{\alpha\dot\alpha\,D}^A + \mathrm{i}\, \overline{F}_{DE} \overline{N}_{\alpha\dot\alpha}^{ECBA} \right.
$$

$$
+ 2 F_{\alpha D} \overline{W}_{\dot\alpha}^{CBA} - 6\, \delta_D^C F_{\alpha E} \overline{W}_{\dot\alpha}^{EBA}
$$

$$
\left. - 3 F^{CE} W_{\alpha EDF} \overline{W}_{\dot\alpha}^{FBA} \right). \tag{13.35}
$$

From $\{\mathcal{D}_{\dot\beta D}, \mathcal{D}_{\dot\alpha C}\} F^{BA}$ one obtains

$$
\sum_{DC} \sum_{BA} \left(8\, \delta_D^B F_C^A + 4\, \delta_D^B \left(F^{AE} \overline{S}_{EC} + \overline{F}_{CE} S^{EA} \right) \right.
$$

$$
+ 2 F^{BE} M_{EDC}^A + 2 \overline{F}_{DE} \overline{M}_C^{EBA} + W_{DCE}^\alpha F_\alpha^{EBA}
$$

$$
\left. + \overline{W}_{\dot\alpha}^{BAE} \overline{F}_{EDC}^{\dot\alpha} + [F^{BA}, \overline{F}_{DC}] \right) = 0. \tag{13.36}
$$

For $N \geq 4$, this equation has a completely traceless part without any linear term:

$$
\sum_{DC} \sum_{BA} \mathrm{tl} \left(2 F^{BE} M_{EDC}^A + 2 \overline{F}_{DE} \overline{M}_C^{EBA} + W_{DCE}^\alpha F_\alpha^{EBA} \right.
$$

$$
\left. + \overline{W}_{\dot\alpha}^{BAE} \overline{F}_{EDC}^{\dot\alpha} + [F^{BA}, \overline{F}_{DC}] \right) = 0. \tag{13.37}
$$

In complete analogy to supergravity (Chapter 6) it can be shown that (13.37) is the consistency condition for the constraints (13.21). This implies that the constraints are consistent for $N \leq 3$ and inconsistent for $N \geq 5$. The case $N = 4$ is special and will be considered in the next section.

In the remainder of this section we shall assume $N \leq 4$ and we shall continue in the linear approximation. In particular, this means that all fields are now abelian. Observe, however, that an abelian F may be rescaled such that the scalars F^{BA} are dimensionless. Therefore products of scalar fields must not be omitted in this approximation. For example, (13.36) becomes

$$F_A^A = 0 \qquad\qquad (N = 2),$$

$$F_B^A = -\frac{1}{2}\left(F^{AC}\,\overline{S}_{CB} + \overline{F}_{BC}\,S^{CA}\right)$$
$$+ \frac{1}{4}\left(F^{CD} M_{BCD}^A + \overline{F}_{CD}\,\overline{M}_B^{ACD}\right) \qquad (N = 3, 4). \qquad (13.38)$$

dim $\frac{5}{2}$

At dimension $\frac{5}{2}$ one obtains from $\{D_\gamma^E, D_\beta^D\} F_\alpha^{CBA}$

$$D_\gamma^E F_{\beta\alpha}^{DCBA} = -\frac{1}{6} \sum_{DCBA} F^{ED} V_{\gamma\beta\alpha}^{CBA}$$

$$+ \frac{1}{12} \sum_{\beta\alpha} \sum_{DCBA} \varepsilon_{\gamma\beta}\left(2\,F^{DC} D_\alpha^B S^{AE} + \frac{5}{3}\mathrm{i}\,F^{ED} \partial_{\alpha\dot\alpha} \overline{W}^{\dot\alpha CBA}\right) \qquad (13.39)$$

and from $\{D_\beta^E, D_{\dot\alpha D}\} F_\alpha^{CBA}$

$$D_{\dot\alpha E} F_{\beta\alpha}^{DCBA} = \frac{1}{4!} \sum_{\beta\alpha} \sum_{DCBA}\left(-4\mathrm{i}\,\delta_E^D \partial_{\beta\dot\alpha} F_\alpha^{CBA} + 3\,F^{DC} D_{\dot\alpha E} V_{\beta\alpha}^{BA}\right.$$
$$\left. + 24\,\delta_E^D F^{CB}\,\overline{\Psi}_{\beta\alpha\dot\alpha}^A\right), \qquad (13.40)$$

$$\partial_{\dot\alpha}^\alpha F_\alpha^{CBA} = \mathrm{i} \sum_{CBA}\left(F^{CD}\,\overline{\Lambda}_{\dot\alpha D}^{BA} + 2\,F^{CB}\,\overline{\Psi}_{\dot\alpha}^A\right). \qquad (13.41)$$

The identity $\{D_\beta^C, D_{\dot\alpha}^B\}\,\overline{F}_{\dot\alpha}^A$ is then satisfied and $\{D_\alpha^C, D_{\dot\beta B}\}\,\overline{F}_{\dot\alpha}^A$ yields

$$D_\alpha^C F_B^A = 2\mathrm{i}\,\delta_B^C \partial_{\alpha\dot\alpha} \overline{F}^{\dot\alpha A} - \mathrm{i}\,\delta_B^A \partial_{\alpha\dot\alpha} \overline{F}^{\dot\alpha C} - \frac{1}{2}\,F^{AD} D_\alpha^C \overline{S}_{DB}$$

$$- \frac{1}{2}\,\overline{F}_{BD} D_\alpha^C S^{DA} - 2\,F^{CD} \Lambda_{\alpha DB}^A + F^{AD} \Lambda_{\alpha DB}^C$$

$$- 4\,\delta_B^C F^{AD} \Psi_{\alpha D} + 2\,\delta_B^A F^{CD} \Psi_{\alpha D} + \frac{\mathrm{i}}{2}\,\overline{F}_{BD} \partial_{\alpha\dot\alpha} \overline{W}^{\dot\alpha DCA}. \qquad (13.42)$$

61

Because of (13.38), this implies

$$\partial_{\alpha\dot\alpha} \overline{F}^{\dot\alpha A} = \frac{i}{2} F^{BC} \Lambda^{A}_{\alpha\,BC} - 2i\,F^{AB}\,\Psi_{\alpha B} + \frac{1}{4}\,\overline{F}_{BC}\,\partial_{\alpha\dot\alpha}\overline{W}^{\dot\alpha\,ABC}$$

$$(N = 3, 4). \qquad (13.43)$$

The remaining dim-$\frac{5}{2}$ identities, $\{D_{\dot\beta E}, D_{\dot\alpha D}\}\,F^{CBA}_{\alpha}$ and $\{D_{\dot\gamma C}, D_{\dot\beta B}\}\,\overline{F}^{A}_{\dot\alpha}$, are satisfied on account of the preceding ones.

dim 3

At dimension 3 we apply D^{D}_{α} to (13.41) and find

$$\partial^{\beta}_{\ \dot\alpha}\,F^{DCBA}_{\beta\alpha} = \frac{1}{4}\sum_{DCBA} F^{DC}(\partial^{\beta}_{\ \dot\alpha}V^{BA}_{\beta\alpha} - 2\,\partial^{\dot\beta}_{\alpha}\,\overline{W}^{BA}_{\dot\beta\dot\alpha}). \qquad (13.44)$$

Similarly, the spinor derivatives of (13.43) give

$$\partial^{\alpha\dot\alpha}\,\partial_{\alpha\dot\alpha}\,F^{BA} = F^{CD}\,P^{BA}_{CD} + \frac{1}{3}\,F^{BA}\,R - \frac{1}{4}\,\overline{F}_{CD}\,\partial^{\alpha\dot\alpha}\,\overline{N}^{BACD}_{\alpha\dot\alpha}$$

$$(N = 3, 4) \qquad (13.45)$$

and

$$\partial_{\alpha}^{\ \dot\beta}\,\overline{F}_{\dot\beta\dot\alpha} = \frac{1}{4}\,(\overline{F}_{AB}\,\partial_{\alpha}^{\ \dot\beta}\,\overline{W}^{AB}_{\dot\beta\dot\alpha} - F^{AB}\,\partial^{\beta}_{\ \dot\alpha}\,W_{\beta\alpha\,AB})$$

$$(N = 3, 4). \qquad (13.46)$$

All other Ricci identities are then satisfied.

The $\theta = \overline{\theta} = 0$ components of (13.41, 43), (13.45), and (13.46) are the field equations for the Weyl spinors, the scalars, and the vector field, respectively. Note that they appear only for $N \geq 3$. (The remaining field equation (13.44) occurs only for $N = 4$ and will be eliminated in the next section.)

Summary

To summarize, the $N \leq 4$ vector multiplets are given by the $\theta = \overline{\theta} = 0$ components of the superfields

$$A_{m}, \ F^{BA}, \ F^{CBA}_{\alpha}, \ F_{\alpha A}, \ F^{A}_{B}. \qquad (13.47)$$

For $N \leq 2$, the multiplet is off-shell and $F^{A}_{B}|$ is an auxiliary field. For $N = 3$ and $N = 4$, F^{A}_{B} disappears and all fields are restricted by equations of motion. The consequences of this fact for $N = 3, 4$ off-shell supergravity will be discussed at the end of this chapter. In the next section we shall impose a further constraint for $N = 4$, which eliminates half of the components of F^{BA} and relates F^{CBA}_{α} to $F_{\alpha A}$.

13.5 $N = 4$

For $N = 4$, the consistency condition (13.37) becomes

$$\text{tl}\left[F^{AB}, \overline{F}_{CD}\right] = 0. \tag{13.48}$$

We now distinguish two cases. In the *non-abelian* case, this equation does not vanish and one has to impose one more constraint to make the theory consistent. The most general form of this constraint is

$$\Phi \, F^{AB} = \frac{1}{2} \, \varepsilon^{ABCD} \, \overline{F}_{CD}, \tag{13.49}$$

where

$$\Phi \, \overline{\Phi} = 1. \tag{13.50}$$

The superfield Φ must have U(1) weight 4 and it must be super-Weyl invariant in order to leave all the conformal symmetries unbroken. We shall determine its explicit form later. The above constraint implies

$$\left[F^{AB}, \overline{F}_{CD}\right] = -\frac{1}{2} \sum_{AB} \sum_{CD} \delta^A_C \left[F^{BE}, \overline{F}_{ED}\right], \tag{13.51}$$

$$\left[F^{AB}, \overline{F}_{AB}\right] = 0. \tag{13.52}$$

Hence the condition (13.48) is identically satisfied. Observe, however, that this does not yet mean that the constraint (13.49) is fully consistent.

In the *abelian* case, the condition (13.48) vanishes identically and the constraints (13.21) are therefore perfectly consistent. Nevertheless, the additional constraint (13.49) may be imposed without loss of generality since it is always possible to decompose F into $\widetilde{F} + \widehat{F}$, where $\widetilde{F}^{AB}$ is self-dual and $\widehat{F}^{AB}$ is anti-self-dual.

It remains to be shown that the constraint (13.49) is actually consistent. To this aim, we first define

$$\overline{W}^{ABC}_{\dot\alpha} = \varepsilon^{ABCD} \, \overline{W}_{\dot\alpha D},$$

$$\overline{W}^{AB}_{\dot\alpha\dot\beta} = \frac{1}{2} \varepsilon^{ABCD} \, \overline{W}_{\dot\alpha\dot\beta CD},$$

$$\overline{M}^{ABC}_D = \varepsilon^{ABCE} \, \overline{M}_{(ED)},$$

$$\overline{N}^{ABCD}_{\alpha\dot\alpha} = \varepsilon^{ABCD} \, \overline{N}_{\alpha\dot\alpha}. \tag{13.53}$$

Then we insert (13.49) into (13.22, 24) and find

$$\mathcal{D}^A_\alpha \Phi = -W^A_\alpha, \tag{13.54}$$

$$F^{ABC}_\alpha = \varepsilon^{ABCD} \left(2 \, \overline{\Phi} \, F_{\alpha D} - \overline{\Phi}^2 \, \overline{F}_{DE} \, W^E_\alpha\right). \tag{13.55}$$

Analogously, a comparison of (13.27) and (13.34) gives

$$F_{\alpha\beta}^{ABCD} = \varepsilon^{ABCD} \, \overline{\Phi} \left(4 \, F_{\alpha\beta} + \frac{1}{2} \, \overline{F}_{EF} \, V_{\alpha\beta}^{EF} - F^{EF} \, W_{\alpha\beta\,EF} \right.$$

$$\left. + \, \overline{\Phi} \sum_{\alpha\beta} F_{\alpha E} \, W_{\beta}^{E} + \overline{\Phi}^{\,2} \, \overline{F}_{EF} \, W_{\alpha}^{E} \, W_{\beta}^{F} \right). \qquad (13.56)$$

We have checked that these are all the consequences of (13.49) that follow from the Bianchi and Ricci identities. Hence the constraint (13.49) is consistent.

*

Finally, we are going to determine the superfield Φ introduced in (13.49). Since Φ must be super-Weyl invariant, we conclude from (13.54)

$$\Phi = \Phi(W_i, \overline{W}^i). \qquad (13.57)$$

Using

$$\mathcal{D}_{\alpha}^{A} \, W_i = \varepsilon_{ij} \, \overline{W}^j \, W_{\alpha}^{A},$$

$$\mathcal{D}_{\dot\alpha A} \, W_i = 0, \qquad (13.58)$$

we conclude further that

$$\frac{\partial \Phi}{\partial W_i} \, \varepsilon_{ij} \, \overline{W}^j = -1. \qquad (13.59)$$

Before we solve this differential equation, we make a change of variables and define

$$x = \frac{W_1}{\overline{W}^2}, \qquad y = \frac{W_2}{\overline{W}^1}, \qquad z = \frac{\overline{W}^2}{\overline{W}^1}. \qquad (13.60)$$

The fourth variable would be the product $\overline{W}^1 \overline{W}^2$, but (8.5) implies

$$x + y = (\overline{W}^1 \overline{W}^2)^{-1}.$$

Thus we now have

$$\Phi = \Phi(x, y, z), \qquad (13.61)$$

where x, y, and z are all independent. Equation (13.59) then becomes

$$\frac{\partial \Phi}{\partial x} - \frac{\partial \Phi}{\partial y} = 1 \qquad (13.62)$$

($\varepsilon_{12} = -1$) and the general solution is

$$\Phi = x + f(x + y, z). \qquad (13.63)$$

Next we employ the condition that Φ must have U(1) weight 4, i.e.,

$$\delta_\Lambda \Phi = 4\,\Lambda\,\Phi.$$

This yields the differential equation

$$x\,\frac{\partial\Phi}{\partial x} + y\,\frac{\partial\Phi}{\partial y} = \Phi. \tag{13.64}$$

Inserting (13.63) into (13.64), we find

$$\Phi = x + (x+y)\,g(z). \tag{13.65}$$

The last condition on Φ (13.50) reads

$$\left| x + (x+y)\,g(z) \right|^2 = 1. \tag{13.66}$$

Using $y = -x\,z\bar{z}$ and $x\bar{x} = (z\bar{z})^{-1}$, this equation can be written as

$$\left| 1 + (1 - z\bar{z})\,g(z) \right|^2 = z\bar{z}, \tag{13.67}$$

which is equivalent to

$$\left| \frac{1 + g(z)}{z\,g(z)} \right|^2 = 1. \tag{13.68}$$

From this we conclude

$$\frac{1 + g(z)}{z\,g(z)} = e^{2i\alpha}, \tag{13.69}$$

where α is real. Since α may depend only on z and not on $\bar{z}$, it must be a constant $(0 \leq \alpha < \pi)$. The last equation implies

$$g(z) = \left(e^{2i\alpha} z - 1 \right)^{-1}. \tag{13.70}$$

Inserting this into (13.65), we obtain the final result

$$\Phi = \frac{e^{i\alpha}\,W_1 + e^{-i\alpha}\,W_2}{e^{i\alpha}\,\overline{W}^2 - e^{-i\alpha}\,\overline{W}^1}. \tag{13.71}$$

Thus the transformation laws (and the action) of the $N - 4$ vector multiplet in curved space include a function $\Phi(W_i, \overline{W}^i, \alpha)$ which breaks the global SU(1, 1) invariance and contains an arbitrary angle α $(0 \leq \alpha < \pi)$. If the gauge group is a direct product, there may be one angle for each factor. The same results were obtained in Ref. [54] using quite different methods.

13.6 $N = 3, 4$ Off-Shell Supergravity

The fact that the $N = 3$ and $N = 4$ vector multiplets are on-shell has immediate consequences for the corresponding Poincaré supergravity theories. Before we discuss this, we recall that the $N \leq 4$ on-shell Poincaré multiplets contain $\binom{N}{2}$ abelian vector fields. In superspace these vector fields are described by one-form gauge potentials, i.e., they must be part of the vector multiplets (13.47). Therefore any $N = 3$ or $N = 4$ Poincaré supergravity multiplet must contain at least 3, resp. 6 vector submultiplets.

One might now argue that it is not possible to construct off-shell Poincaré supergravities for $N = 3, 4$ because there are no off-shell vector multiplets. This argument, however, is wrong. In Chapter 15 we shall in fact derive two $N = 2$ off-shell Poincaré theories using on-shell "compensating" multiplets. There are in principle two possibilities to get rid of the field equations:

 (i) absorb the field equations by fields of the conformal supergravity multiplet,

 (ii) interpret the field equations as Bianchi identities of the dual fields.

For fermions, only the first possibility may work. Hence the field equations (13.41) for $F_{\alpha i}^{CBA}$ and (13.43) for $F_{\alpha A}^i$ must be absorbed by $\overline{\Lambda}_{\dot\alpha C}^{BA}$ ($i = 1, \ldots, n$; $n = $ number of vector multiplets). However, for $N = 3$ we have at least 12 field equations while $\overline{\Lambda}$ has only 9 components, and for $N = 4$ there are at least 24 field equations whereas $\overline{\Lambda}$ has only 20 components. From this we conclude that the $N = 3$ and $N = 4$ Poincaré supergravities have no off-shell formulation in conventional extended superspace.

14. Two-Form Gauge Potentials

14.1 Superspace Geometry

Another important off-shell multiplet appearing in Poincaré supergravity theories is the tensor multiplet. In superspace it is described by a real two-form gauge potential

$$B = \frac{1}{2} dz^M dz^N B_{NM} = \frac{1}{2} E^A E^B B_{BA} \tag{14.1}$$

with the transformation law

$$\delta B = d\Omega. \tag{14.2}$$

The invariant field strength

$$G = dB, \tag{14.3}$$

$$G = \frac{1}{3!} dz^M dz^N dz^L G_{LNM} = \frac{1}{3!} E^A E^B E^C G_{CBA} \tag{14.4}$$

is a real 3-form and satisfies the Bianchi identity

$$dG = 0. \tag{14.5}$$

The structure equation (14.3) reads explicitly

$$G_{LNM} = \frac{1}{2} \sum_{LNM} \partial_L B_{NM}, \tag{14.6}$$

$$G_{CBA} = \frac{1}{2} \sum_{CBA} \left(\mathcal{D}_C B_{BA} + T_{CB}{}^D B_{DA} \right) \tag{14.7}$$

and the Bianchi identity (14.5) can be written as

$$\sum_{DCBA} \left(\mathcal{D}_D G_{CBA} + \frac{3}{2} T_{DC}{}^{\mathcal{E}} G_{\mathcal{E}BA} \right) = 0. \tag{14.8}$$

The transformation law (14.2) is equivalent to

$$\delta B_{NM} = \xi^L \partial_L B_{NM} + \sum_{NM} \left[(\partial_N \xi^L) B_{LM} + \partial_N \Omega_M \right]. \tag{14.9}$$

Defining

$$\omega_M = \Omega_M + \xi^N B_{NM}, \tag{14.10}$$

one obtains the more covariant form

$$\delta B_{NM} = \xi^A G_{ANM} + \sum_{NM} \partial_N \omega_M. \tag{14.11}$$

14.2 Constraints

The lowest-dimensional components of the field strength G are $G_{\gamma\beta\alpha}^{CBA}$, $G_{\gamma\beta A}^{CB\dot{\alpha}}$, and their complex conjugates. The component $B_{\beta a}^{B}$ of the two-form gauge potential has the same dimension and we may therefore impose a conventional constraint which eliminates $B_{\beta a}^{B}$ as an independent variable. From (14.7) we find that this constraint is

$$G_{\beta\alpha\,\dot{\alpha}A}^{BA} = 0. \qquad (14.12)$$

All the remaining parts of $G_{\gamma\beta\alpha}^{CBA}$ and $G_{\gamma\beta A}^{CB\dot{\alpha}}$ have high spins, so we have to set them equal to zero, too. Thus the dim-$\frac{3}{2}$ constraints are simply

$$G_{\underline{\gamma\beta\alpha}} = 0. \qquad (14.13)$$

At dimension 2 we can use B_{ba} to absorb parts of $G_{\gamma\beta a}^{CB}$ and $G_{\gamma B a}^{C\dot{\beta}}$. The corresponding conventional constraint is

$$\sum_{\beta\alpha} G_{\beta A\,\alpha\dot{\alpha}}^{A\dot{\alpha}} = 0. \qquad (14.14)$$

For $N = 1$, the multiplet defined by the above constraints is still reducible and we shall impose one more constraint in the next section. For $N = 2$, the constraints (14.13–14) lead to an irreducible off-shell multiplet. For higher N, however, we shall find that the field strength G vanishes identically. This agrees with the well-known fact that there are no scalar multiplets for $N > 2$.

14.3 Bianchi Identities

We are now going to solve the Bianchi identities (14.8) subject to the constraints (14.13) and (14.14).

dim 2: $\left(\genfrac{}{}{0pt}{}{}{\underline{\delta\gamma\beta\alpha}}\right)$

At dimension 2, the identity $\left(\genfrac{}{}{0pt}{}{DCBA}{\delta\gamma\beta\alpha}\right)$ is satisfied and $\left(\genfrac{}{}{0pt}{}{DCB\dot{\alpha}}{\delta\gamma\beta A}\right)$ gives

$$G_{\gamma\beta\,\alpha\dot{\alpha}} = i\sum_{\gamma\beta}\varepsilon_{\gamma\alpha}\,G_{\beta\dot{\alpha}} \qquad (N=1),$$

$$G_{\gamma\beta\,\alpha\dot{\alpha}}^{CB} = 0 \qquad\qquad (N\geq 2). \qquad (14.15)$$

For $N = 1$, $G_{\alpha\dot{\alpha}}$ would describe a physical vector field. Since we want to end up with an irreducible scalar multiplet, we impose the additional constraint

$$G_{\alpha\dot{\alpha}} = 0 \qquad (N=1). \qquad (14.16)$$

The last dim-2 identity $\left(\begin{smallmatrix} DC\dot\beta\dot\alpha \\ \delta\gamma BA \end{smallmatrix}\right)$ together with the constraint (14.14) yields

$$G^{C}_{\gamma\dot\beta B\,\alpha\dot\alpha} = i\,\varepsilon_{\gamma\alpha}\,\varepsilon_{\dot\beta\dot\alpha}\,G^{C}_{B}\,,\tag{14.17}$$

where

$$G^{A}_{A} = 0 \qquad (N = 2),$$
$$G^{A}_{B} = 0 \qquad (N \geq 3).\tag{14.18}$$

dim $\frac{5}{2}$: $\left(\begin{smallmatrix} \\ \delta\gamma\beta a \end{smallmatrix}\right)$

At dimension $\frac{5}{2}$, $\left(\begin{smallmatrix} DCB \\ \delta\gamma\beta a \end{smallmatrix}\right)$ is identically satisfied and $\left(\begin{smallmatrix} DC\dot\beta \\ \delta\gamma B a \end{smallmatrix}\right)$ implies

$$G^{C}_{\gamma\,\beta\dot\beta\,\alpha\dot\alpha} = \varepsilon_{\dot\beta\dot\alpha}\sum_{\beta\alpha}\varepsilon_{\gamma\beta}\,G^{C}_{\alpha}\,,\tag{14.19}$$

$$\mathcal{D}^{C}_{\alpha}\,G^{A}_{B} = 4\,\delta^{C}_{B}\,G^{A}_{\alpha} - 2\,\delta^{A}_{B}\,G^{C}_{\alpha}\,.\tag{14.20}$$

dim 3: $\left(\begin{smallmatrix} \\ \delta\gamma\,ba \end{smallmatrix}\right)$

The first dim-3 identity, $\left(\begin{smallmatrix} DC \\ \delta\gamma\,ba \end{smallmatrix}\right)$, gives

$$\mathcal{D}^{B}_{\beta}\,G^{A}_{\alpha} = \varepsilon_{\beta\alpha}\,G^{[BA]} - \frac{1}{2}\sum_{BA}\widehat{G^{B}_{C}\left(V^{CA}_{\beta\alpha} + \varepsilon_{\beta\alpha}\,S^{CA}\right)}.\tag{14.21}$$

Next we define

$$G_{cba} = \varepsilon_{cbad}\,\widetilde{G}^{d}\,,\tag{14.22}$$

which is equivalent to

$$G_{\gamma\dot\gamma\,\beta\dot\beta\,\alpha\dot\alpha} = 2i\left(\varepsilon_{\gamma\alpha}\,\varepsilon_{\dot\gamma\dot\beta}\,\widetilde{G}_{\beta\dot\alpha} - \varepsilon_{\gamma\beta}\,\varepsilon_{\dot\gamma\dot\alpha}\,\widetilde{G}_{\alpha\dot\beta}\right).\tag{14.23}$$

From $\left(\begin{smallmatrix} D\dot\gamma \\ \delta C\,ba \end{smallmatrix}\right)$ we then obtain

$$\mathcal{D}_{\dot\alpha B}\,G^{A}_{\alpha} = -2\,\delta^{A}_{D}\,\widetilde{G}_{\alpha\dot\alpha} - \frac{i}{2}\,\mathcal{D}_{\alpha\dot\alpha}\,G^{A}_{B} + G^{A}_{C}\,U^{C}_{\alpha\dot\alpha\,B}\,.\tag{14.24}$$

As we anticipated in the preceding section, the above equations imply

$$G_{CBA} = 0 \qquad (N \geq 3).\tag{14.25}$$

dim $\frac{7}{2}$: $\left(\underline{\delta}\,cba\right)$

The only identity at dimension $\frac{7}{2}$ reads

$$\mathcal{D}^D_\delta G_{cba} = \frac{1}{2} \sum_{cba} \left(\mathcal{D}_c G^D_{\delta ba} + T_{c\delta}^{D\varepsilon} G_{\underline{\varepsilon}\,ba} - T_{cb\dot{\varepsilon}}^{\phantom{cb\dot{\varepsilon}}E} G^{D\dot{\varepsilon}}_{\delta E a} \right), \tag{14.26}$$

which is equivalent to

$$
\begin{aligned}
\mathcal{D}^A_\beta \tilde{G}_{\alpha\dot{\alpha}} &= \mathrm{i}\, \mathcal{D}_{\beta\dot{\alpha}} G^A_\alpha - \frac{\mathrm{i}}{2}\, \mathcal{D}_{\alpha\dot{\alpha}} G^A_\beta - \frac{1}{2}\, G^A_B\, \overline{\Psi}^B_{\beta\alpha\dot{\alpha}} \\
&\quad + \frac{3}{2}\, V^{AB}_{\beta\alpha}\, \overline{G}_{\dot{\alpha}B} + \frac{3}{2}\, U^A_{\beta\dot{\alpha}B}\, G^B_\alpha \\
&\quad + \frac{1}{2}\, \varepsilon_{\beta\alpha} \left(3\, G^A_B\, \overline{\Psi}^B_{\dot{\alpha}} + \overline{W}^{AB}_{\dot{\alpha}\dot{\beta}}\, \overline{G}^{\dot{\beta}}_B - 3\, S^{AB}\, \overline{G}_{\dot{\alpha}B} \right).
\end{aligned}
\tag{14.27}
$$

dim 4: $\left(dcba\right)$

Finally, we are left with the identity

$$\sum_{dcba} \left(\mathcal{D}_d G_{cba} + \frac{3}{2}\, T_{dc}^{\varepsilon}\, G_{\underline{\varepsilon}\,ba} \right) = 0, \tag{14.28}$$

which gives

$$\mathcal{D}^{\alpha\dot{\alpha}} \tilde{G}_{\alpha\dot{\alpha}} - 3\mathrm{i} \left(\Psi^\alpha_A\, G^A_\alpha - \overline{\Psi}^A_{\dot{\alpha}}\, \overline{G}^{\dot{\alpha}}_A \right) = 0. \tag{14.29}$$

14.4 Ricci Identities

We still have to solve the Ricci identities for the various superfields introduced in the preceding section. Fortunately, only two of them are non-trivial. The first one is $\{\mathcal{D}^C_\gamma, \mathcal{D}^B_\beta\}\, G^A_\alpha$, which implies

$$\mathcal{D}^C_\alpha G^{BA} = 2\, G^{\beta C}\, V^{BA}_{\beta\alpha} + \frac{1}{3} \sum_{BA} G^C_D\, \mathcal{D}^B_\alpha S^{AD}, \tag{14.30}$$

and the second one is $\{\mathcal{D}^C_\beta, \mathcal{D}_{\dot{\alpha}B}\}\, G^A_\alpha$ yielding

$$
\begin{aligned}
\mathcal{D}_{\dot{\alpha}C}\, G^{BA} &= \sum_{BA} \Big(-2\mathrm{i}\, \delta^B_C\, \mathcal{D}^\alpha_{\dot{\alpha}}\, G^A_\alpha + \frac{1}{2}\, G^B_D\, \mathcal{D}_{\dot{\alpha}C}\, S^{AD} \\
&\quad + \frac{1}{2}\, G^D_C\, \overline{\Lambda}^{BA}_{\dot{\alpha}D} + 4\, G^B_C\, \overline{\Psi}^A_{\dot{\alpha}} + \overline{W}^{BA}_{\dot{\alpha}\dot{\beta}}\, \overline{G}^{\dot{\beta}}_C \\
&\quad + 2\, \delta^B_C\, S^{AD}\, \overline{G}_{\dot{\alpha}D} + 2\, G^{\alpha B}\, U^A_{\alpha\dot{\alpha}C} \Big).
\end{aligned}
\tag{14.31}
$$

All other Ricci identities are then satisfied. Hence the constraints (14.13–14) are consistent.

To summarize, the $N = 1$ and $N = 2$ tensor multiplets are given by the $\theta = \overline{\theta} = 0$ components of the superfields

$$B_{mn}, \quad G^A_B, \quad G^A_\alpha, \quad G^{BA}. \tag{14.32}$$

The multiplets are off-shell and $G^{BA}|$ is an auxiliary field (for $N = 2$). On-shell the antisymmetric tensor is equivalent to a real scalar field [55].

14.5 Solution of the Constraints

In order to construct invariant actions for the $N = 1$ and $N = 2$ tensor multiplets, we have to solve the constraints (14.13–14) in terms of unconstrained or at least chiral prepotentials. This can easily be done using the close relationship between vector and tensor multiplets. The gauge potential A corresponds to the parameter Ω, the field strength F to the potential B, and the Bianchi identity $\mathcal{D}F$ to the field strength G. Therefore the prepotentials of the tensor multiplets should be given by the basic chiral superfields of the vector multiplets. This is confirmed by the fact that the field content of a chiral superfield F_α $(N = 1)$, resp. $\overline{F}$ $(N = 2)$ is just a vector multiplet plus a tensor multiplet.

To proceed in a systematic way, we restrict the dim-1 components of the two-form potential by

$$B^{BA}_{\beta\alpha} = \varepsilon_{\beta\alpha} B^{[BA]},$$

$$B^B_{\beta\dot{\alpha}A} = 0,$$

$$B_{\dot{\beta}B\dot{\alpha}A} = -\varepsilon_{\dot{\beta}\dot{\alpha}} \overline{B}_{[BA]} \tag{14.33}$$

and solve the structure equation (14.7) subject to these "constraints". In the following we assume $N \leq 2$.

At dimension $\frac{3}{2}$ one finds

$$B^B_{\beta\alpha\dot{\alpha}} = i\varepsilon_{\beta\alpha} \overline{B}^B_{\dot{\alpha}} \tag{14.34}$$

and the conditions

$$\mathcal{D}^C_\alpha B^{BA} = 0, \tag{14.35}$$

$$\mathcal{D}_{\dot{\alpha}C} B^{BA} = 2 \sum_{BA} \delta^B_C \overline{B}^A_{\dot{\alpha}}. \tag{14.36}$$

At dimension 2, the structure equation yields

$$B_{\beta\dot{\beta}\alpha\dot{\alpha}} = \varepsilon_{\dot{\beta}\dot{\alpha}} B_{(\beta\alpha)} + \varepsilon_{\beta\alpha} \overline{B}_{(\dot{\beta}\dot{\alpha})} \tag{14.37}$$

and the conditions

$$\mathcal{D}_\alpha^B \, \overline{B}_{\dot{\alpha}}^A = -\mathrm{i}\, \mathcal{D}_{\alpha\dot{\alpha}} \, B^{BA} + B^{AC} \, U_{\alpha\dot{\alpha}C}^B \,, \tag{14.38}$$

$$\mathcal{D}_{\dot{\beta}B} \, \overline{B}_{\dot{\alpha}}^A = -2\,\delta_B^A \, \overline{B}_{\dot{\beta}\dot{\alpha}} - B^{AC} \, \overline{V}_{\dot{\beta}\dot{\alpha}CB} - \overline{B}_{BC} \, \overline{W}_{\dot{\beta}\dot{\alpha}}^{CA}$$
$$+ \frac{1}{2}\,\varepsilon_{\dot{\beta}\dot{\alpha}} \,(G_B^A + \mathrm{i}\, B_B^A - B^{AC} \, \overline{S}_{CB} - \overline{B}_{BC} \, S^{CA})\,, \tag{14.39}$$

where

$$B_B^A = \overline{B}_B^A \,. \tag{14.40}$$

In addition, one obtains from the Ricci identity $\{\mathcal{D}_{\dot{\beta}D}, \mathcal{D}_{\dot{\alpha}C}\}\, B^{BA}$:

$$B_A^A = 0 \qquad (N = 2). \tag{14.41}$$

The solution of the dim-$\frac{5}{2}$ and dim-3 identities is completely analogous to the calculations that we have done in Sections 13.3 and 13.4. We skip these identities here.

The transformation law (14.2) of the two-form gauge potential can now be written as

$$\delta B = \mathrm{i}\, F \,, \tag{14.42}$$

where F is restricted by the constraints (13.21). (F was supposed to be antihermitian.) This transformation law allows us to choose the gauge

$$B^{BA}\Big|_{\theta=\bar{\theta}=0} = 0 \,,$$

$$\overline{B}_{\dot{\alpha}}^A\Big| = 0 \,,$$

$$B_B^A\Big| = 0 \,. \tag{14.43}$$

The remaining component fields of B^{BA} form exactly the tensor multiplet (14.32).

15. $N = 2$ Off-Shell Supergravity

We now continue our discussion of off-shell Poincaré supergravity with the case $N = 2$. In the following we shall use the definitions

$$
\begin{aligned}
V^{BA}_{\beta\alpha} &= \varepsilon^{BA} V_{\beta\alpha}, \\
\overline{W}^{BA}_{\dot\beta\dot\alpha} &= \varepsilon^{BA} \overline{W}_{\dot\beta\dot\alpha}, \\
\overline{\Lambda}^{BA}_{\dot\alpha C} &= \varepsilon^{BA} \overline{\Lambda}_{\dot\alpha C}, \\
P^{BA}_{DC} &= \varepsilon^{BA} \varepsilon_{DC} P.
\end{aligned}
\tag{15.1}
$$

15.1 The Minimal Field Representation

As we already mentioned in Section 10.1, the most convenient starting point for the derivation of $N \leq 4$ Poincaré supergravities is the corresponding conformal theories. The two steps of the construction are:

 (i) couple conformal supergravity to matter multiplets,

 (ii) break conformal supersymmetry.

We shall see that this mechanism works almost automatically in superspace and that all the minimal off-shell Poincaré supergravities can be derived this way.

For $N = 2$, one of the matter multiplets must be an abelian vector multiplet described by

$$
\begin{aligned}
F^{BA} &= \varepsilon^{BA} F, \\
\overline{F}_{BA} &= -\varepsilon_{BA} \overline{F}.
\end{aligned}
\tag{15.2}
$$

$\overline{F}$ is a chiral superfield with U(1) weight 2 and it is Weyl-covariant with Weyl weight 1. Hence [1]

$$
\mathcal{S} = \int \mathrm{d}^4 x \, \mathrm{d}^4\Theta \; \mathcal{E} \, \overline{F} \, \overline{F} + \text{c.c.}
\tag{15.3}
$$

is a super-Weyl invariant chiral action for the vector multiplet.

Now we break the super-Weyl invariance and the local U(1) symmetry by choosing the gauge

$$
F = \overline{F} = 1.
\tag{15.4}
$$

The action (15.3) becomes then simply

$$
\mathcal{S} = \int \mathrm{d}^4 x \, \mathrm{d}^4\Theta \; \mathcal{E} + \text{c.c.}.
\tag{15.5}
$$

---

[1]   Here and in the following we assume that all superfields in chiral actions depend only on the chiral coordinates (x, Θ).

The consequences of the constraint (15.4) can easily be obtained from the solution of the various Bianchi and Ricci identities. At dimension $\frac{1}{2}$, (13.22) and (13.24) yield

$$\Phi_\alpha{}^A = 0, \tag{15.6}$$

$$\overline{F}_{\dot\alpha}{}^A = 0 \tag{15.7}$$

and at dimension 1 we find

$$U^B{}_{\alpha\dot\alpha A} = \delta^B_A\, U_{\alpha\dot\alpha}, \tag{15.8}$$

$$\Phi_{\alpha\dot\alpha} = -\frac{i}{2}\, U_{\alpha\dot\alpha}, \tag{15.9}$$

and

$$\overline{F}_{\dot\beta\dot\alpha} = \frac{1}{2}\left(\overline{W}_{\dot\beta\dot\alpha} - \overline{V}_{\dot\beta\dot\alpha}\right), \tag{15.10}$$

$$F^A{}_B = 0, \tag{15.11}$$

$$S^{BA} = -\overline{S}^{BA}. \tag{15.12}$$

In the following we shall keep the full $SL(2, \mathbf{C}) \times U(2)$ covariant derivatives, although the $U(1)$ connection is now given by (15.6,9). This causes no problems except that the $U(1)$ weights can no longer be read off from the internal indices (ε^{AB} has $U(1)$ weight 0).

The above equations imply at dimension $\frac{3}{2}$

$$\mathcal{D}^A_\alpha \overline{W}_{\dot\beta\dot\alpha} = 0,$$

$$\mathcal{D}_{\dot\gamma A} \overline{W}_{\dot\beta\dot\alpha} = 2\,\overline{W}_{\dot\gamma\dot\beta\dot\alpha A} + \sum_{\dot\beta\dot\alpha} \varepsilon_{\dot\gamma\dot\beta}\, \overline{A}_{\dot\alpha A}, \tag{15.13}$$

$$\mathcal{D}^A_\gamma V_{\beta\alpha} = \sum_{\beta\alpha} \varepsilon_{\gamma\beta}\left(\Lambda^A_\alpha - 2\,\Psi^A_\alpha\right),$$

$$\mathcal{D}_{\dot\alpha A} V_{\beta\alpha} = 2\,\overline{\Psi}_{\beta\alpha\dot\alpha A}, \tag{15.14}$$

$$\mathcal{D}_{\dot\alpha C} S^{BA} = -\sum_{BA} \widehat{\delta}^B_C\left(\overline{\Lambda}^A_{\dot\alpha} + 2\,\overline{\Psi}^A_{\dot\alpha}\right), \tag{15.15}$$

$$\mathcal{D}^A_\beta U_{\alpha\dot\alpha} = -\overline{\Psi}^A_{\beta\alpha\dot\alpha} - \varepsilon_{\beta\alpha}\left(\overline{\Psi}^A_{\dot\alpha} + 2\,\overline{\Lambda}^A_{\dot\alpha}\right) \tag{15.16}$$

and at dimension 2

$$\rho^A_{\beta\alpha A} = -\frac{i}{2} \sum_{\beta\alpha} \mathcal{D}_\beta{}^{\dot\alpha} U_{\alpha\dot\alpha}, \tag{15.17}$$

$$\mathcal{D}^\beta{}_{\dot\alpha} F_{\beta\alpha} + \mathcal{D}_\alpha{}^{\dot\beta} \overline{F}_{\dot\beta\dot\alpha} = 0. \tag{15.18}$$

These are all the superfield equations that follow from the constraint (15.4).

In order to find the independent component fields, we first consider the $\theta = \bar{\theta} = 0$ components of the vielbein, SU(2) connection, and vector potential. The transformation laws (2.33–34) and (13.17) allow us to choose the gauge

$$
E_{\mathcal{M}}{}^{\mathcal{A}}\big|_{\theta=\bar{\theta}=0} = \begin{pmatrix} e_m{}^a & \frac{1}{2}\psi_m{}^{\underline{\alpha}} \\ 0 & \delta^{\underline{\alpha}}_{\underline{\mu}} \end{pmatrix},
\tag{15.19}
$$

$$
\widetilde{\Phi}_{\mathcal{M}}{}^{B}{}_{A}\big| = (\,i\,v_{mA}^{B}\,,\,0\,),
\tag{15.20}
$$

$$
A_{\mathcal{M}}\big| = (\,i\,a_m\,,\,0\,).
\tag{15.21}
$$

The remaining component fields are simply the $\theta = \bar{\theta} = 0$ components of the independent covariant superfields. Altogether, we obtain the *minimal field representation* [56]

$$
e_m{}^a,\ \psi_{mA}{}^{\alpha},\ a_m,\ v_{mA}^{B},\ V_{\beta\alpha},\ S^{BA},\ U_{\alpha\dot\alpha},\ \overline{\Lambda}_{\dot{A}A},\ P
$$
$$
(\theta = \bar{\theta} = 0)
\tag{15.22}
$$

with 32 bosonic and 32 fermionic degrees of freedom. All the different $N = 2$ off-shell Poincaré multiplets can be derived from this field representation by breaking the local SU(2) symmetry.

15.2 Multiplets Without Local SO(2)

Minimal 40 + 40 Multiplet. The first possibility is to break the SU(2) completely. We shall do this by imposing a constraint on the SU(2) connection which restricts the SU(2) parameter $\widetilde{\Lambda}^{B}{}_{A}$ to a minimal off-shell multiplet – the tensor multiplet (14.32). The transformation law

$$
\delta_\Lambda\,\widetilde{\Phi}_c{}^{B}{}_{A} = -\mathcal{D}_c\,\widetilde{\Lambda}^{B}{}_{A} - \Lambda_c{}^{D}\,\widetilde{\Phi}_{D}{}^{B}{}_{A}
\tag{15.23}
$$

shows that this constraint must be [1]

$$
\sum_{CBA}\widehat{\ } \ \widetilde{\Phi}_\gamma^{CBA} = 0.
\tag{15.24}
$$

The SU(2) connection thus becomes

$$
\widetilde{\Phi}_\gamma^{CBA} = \sum_{BA}\widehat{\ } \ \varepsilon^{CB}\,\varphi_\gamma^{A},
$$

---

[1]   Actually, this constraint restricts $\widetilde{\Lambda}^{B}{}_{A}$ to a so-called non-linear multiplet [40], which differs from an ordinary tensor multiplet by non-linear terms in the transformation laws.

$$\widetilde{\varPhi}_{\dot\gamma}^{\ CBA} = \sum_{BA} \varepsilon^{CB}\, \overline{\varphi}_{\dot\gamma}^{\ A}\,, \tag{15.25}$$

whereby φ_α^A is defined.

The consequences of the constraint (15.24) can be derived from the structure equation

$$\widetilde{R}_{DC}{}^{B}{}_{A} = \sum_{DC}\big(\mathcal{D}_D\,\widetilde{\varPhi}_C{}^{B}{}_A + \widetilde{\varPhi}_D{}^{B}{}_E\,\widetilde{\varPhi}_C{}^{E}{}_A\big) + T_{DC}{}^{\varepsilon}\,\widetilde{\varPhi}_\varepsilon{}^{B}{}_A\,. \tag{15.26}$$

($\mathcal{D}$ is still the full $\mathrm{SL}(2,\mathbf{C}) \times \mathrm{U}(2)$ covariant derivative.) From the equation with indices $\binom{DC}{\delta\,\gamma}$ one obtains

$$\mathcal{D}_\beta^B\,\varphi_\alpha^A = -\,\varepsilon^{BA}\big(V_{\beta\alpha} + \varphi_{\beta C}\,\varphi_\alpha^C\big)$$
$$+\,\varepsilon_{\beta\alpha}\big(\varepsilon^{BA}\,\varPhi + S^{BA} + \varphi^{\gamma B}\,\varphi_\gamma^A\big) \tag{15.27}$$

and $\binom{D\,\dot\gamma}{\delta\,C}$ gives

$$\mathcal{D}_{\dot\alpha B}\,\varphi_\alpha^A = \mathrm{i}\,\widetilde{\varPhi}_{\alpha\dot\alpha\,B}^{\ A} - \delta_B^A\big(U_{\alpha\dot\alpha} + \mathrm{i}\,\varphi_{\alpha\dot\alpha}\big) - 2\,\varphi_{\alpha B}\,\overline{\varphi}_{\dot\alpha}^{\ A}\,, \tag{15.28}$$

where

$$\varphi_{\alpha\dot\alpha} = \overline{\varphi}_{\alpha\dot\alpha}\,. \tag{15.29}$$

Further consequences are at dimension $\frac{3}{2}$ (in the linear approximation)

$$D_\beta^C\,\widetilde{\varPhi}_{\alpha\dot\alpha}^{\ BA} = -\mathrm{i}\sum_{BA}\varepsilon^{CB}\Big[\,\overline{\varPsi}_{\beta\alpha\,\dot\alpha}^{\ A} + \varepsilon_{\beta\alpha}\big(\overline{\varPsi}_{\dot\alpha}^{\ A} - \overline{\varLambda}_{\dot\alpha}^{A}\big) + \mathrm{i}\,\partial_{\alpha\dot\alpha}\,\varphi_\beta^A\,\Big]\,, \tag{15.30}$$

$$D_\alpha^A\,\varPhi = 0\,,$$

$$D_{\dot\alpha A}\,\varPhi = \overline{\varLambda}_{\dot\alpha A} - 4\,\overline{\varPsi}_{\dot\alpha A} - 2\mathrm{i}\,\partial_{\alpha\dot\alpha}\,\varphi_A^\alpha\,, \tag{15.31}$$

$$D_\beta^A\,\varphi_{\alpha\dot\alpha} = -\mathrm{i}\,\varepsilon_{\beta\alpha}\big(\overline{\varLambda}_{\dot\alpha}^{A} - 4\,\overline{\varPsi}_{\dot\alpha}^{A}\big) + 2\,\partial_{\beta\dot\alpha}\,\varphi_\alpha^A - \partial_{\alpha\dot\alpha}\,\varphi_\beta^A \tag{15.32}$$

and at dimension 2

$$\widetilde{\rho}_{\beta\alpha\,A}^{\ B} = \frac{1}{2}\sum_{\beta\alpha}\partial_\beta{}^{\dot\alpha}\,\widetilde{\varPhi}_{\alpha\dot\alpha\,A}^{\ B}\,, \tag{15.33}$$

$$P = -\frac{1}{3}\,R - \partial^{\alpha\dot\alpha}\,\varphi_{\alpha\dot\alpha}\,. \tag{15.34}$$

The independent component fields form the *minimal* $40 + 40$ *multiplet* [39, 6]

$$\left(e_m{}^a,\ \psi_{mA}{}^\alpha,\ a_m \mid \varphi_\alpha^A,\ V_{\beta\alpha},\ S^{BA},\ U_{\alpha\dot\alpha},\ \widetilde{\Phi}_{\alpha\dot\alpha A}^B,\ \Phi,\ \varphi_{\alpha\dot\alpha},\ \overline{\Lambda}_{\dot\alpha A}\right)$$

$$(\theta = \overline\theta = 0). \qquad (15.35)$$

(The vertical bar separates the physical from the auxiliary fields.) The action for this multiplet is given by (15.5).

Minimal $32 + 32$ Multiplet. A surprising property of the multiplet (15.35) is that it can be further reduced by restricting the compensating tensor multiplet to an on-shell scalar multiplet. This reduction generates new physical degrees of freedom, but it does not lead to field equations for the supergravity multiplet.

The additional constraint is simply

$$\Phi = \overline\Phi = 0, \qquad (15.36)$$

since $\Phi|$ corresponds to the auxiliary field of the tensor multiplet. The consequences of this constraint are in the linear approximation

$$\overline\Lambda_{\dot\alpha A} = 4\,\overline\Psi_{\dot\alpha A} + 2\mathrm{i}\,\partial_{\alpha\dot\alpha}\,\varphi_A^\alpha, \qquad (15.37)$$

$$\sum_{\dot\beta\dot\alpha} \partial^\alpha{}_{\dot\beta}\,\varphi_{\alpha\dot\alpha} = 0, \qquad (15.38)$$

$$\partial^{\alpha\dot\alpha}\,\widetilde\Phi_{\alpha\dot\alpha A}^B = 0. \qquad (15.39)$$

Note that these equations correspond exactly to the field equations for the tensor multiplet. However, in (15.37) the field equation is absorbed by $\overline\Lambda_{\dot\alpha A}$ and (15.38–39) can be interpreted as Bianchi identities. This is the mechanism that we have already described in Section 13.6.

To be concrete, (15.38) is the Bianchi identity of a real superfield φ satisfying

$$\mathcal{D}_\alpha^A \varphi = \varphi_\alpha^A, \qquad (15.40)$$

which implies

$$\varphi_{\alpha\dot\alpha} = \mathcal{D}_{\alpha\dot\alpha}\,\varphi. \qquad (15.41)$$

The Bianchi identity (15.39) can be solved by three abelian tensor fields $b_{[mn]A}^B$ (traceless in BA). In superspace they are described by three two-form gauge potentials B_A^B. The field strength

$$G_A^B = \mathrm{d}B_A^B$$

satisfies the usual constraints (14.13–14) and the basic superfield $(G_C^D)_A^B$ is given by

$$(G_C^D)_A^B = (\delta_C^D \, \delta_A^B - 2\, \delta_A^D \, \delta_C^B)\, e^{-2\varphi}.$$ (15.42)

Thus we obtain the *minimal* $32 + 32$ *multiplet* [42]

$$\left(e_m{}^a,\, \psi_{mA}{}^\alpha,\, a_m \mid \varphi_\alpha^A,\, \varphi,\, b_{mnA}^B \mid V_{\beta\alpha},\, S^{BA},\, U_{\alpha\dot\alpha}\right)$$
$$(\theta = \bar\theta = 0).$$ (15.43)

It is off-shell irreducible although the fields φ_α^A, φ, and b_{mnA}^B describe a physical scalar multiplet. The action is again given by (15.5).

15.3 Multiplets with Local SO(2)

The local U(2) symmetry of $N = 2$ conformal supergravity does not have to be broken completely since the gauge fields may appear as auxiliary fields in an off-shell Poincaré multiplet. This is possible because there is an invariant coupling of a tensor and an abelian vector field:

$$S = \int d^4x \; e \, \varepsilon^{klmn}\, b_{kl}\, \partial_m v_n$$ (15.44)

$(e = \det e_m{}^a)$. This term is invariant both under

$$\delta\, b_{mn} = \partial_m \omega_n - \partial_n \omega_m$$

and under

$$\delta v_m = -\partial_m \lambda$$

because of

$$\partial_k \left(e\, \varepsilon^{klmn}\right) = 0.$$

The super-Weyl invariant extension is

$$S = \int d^4x \; d^4\Theta \; \mathcal{E} \; \overline{B}\, F + \text{c.c.},$$ (15.45)

where $\overline{B}$ is the chiral prepotential (14.33) of the $N = 2$ tensor multiplet. One can check explicitly that this action contains the tensor field b_{mn} only in the combination (15.44).

Since the vector field v_m has to be abelian, the only possibilities are a local U(1) or a local SO(2) symmetry. However, the local U(1) can always be broken by the gauge (15.4) (otherwise the multiplet would be reducible). Hence minimal $N = 2$ off-shell supergravity may have at most an extra local SO(2) invariance.

New Minimal $40 + 40$ Multiplet. The local SU(2) can be broken down to SO(2) by the gauge condition

$$G^A_B = i \, \varepsilon^{AB} \, e^\varphi . \tag{15.46}$$

G^A_B is the basic superfield of the tensor multiplet containing b_{mn} and φ is a real superfield. Defining

$$\varphi^A_\alpha = \mathcal{D}^A_\alpha \varphi , \tag{15.47}$$

one then obtains from (14.20)

$$G^A_\alpha = -\frac{i}{2} \, e^\varphi \, \varphi_{\alpha A} , \tag{15.48}$$

$$\tilde{\Phi}^{CB}_{\gamma \ A} = \varepsilon^{BA} \left(\tilde{\Phi}^C_\gamma - \frac{1}{2} \varphi_{\gamma C} \right) + \delta^C_A \varphi^B_\gamma - \frac{1}{2} \delta^B_A \varphi^C_\gamma . \tag{15.49}$$

Observe that, with this definition, the SO(2) connection $\tilde{\Phi}$ is a real 1-form (the SU(2) connection is antihermitian).

At dimension 1, (14.21) yields

$$\mathcal{D}^B_\beta \varphi^A_\alpha = \varepsilon^{BA} \left(2 V_{\beta\alpha} + \frac{1}{2} \varphi_{\beta C} \varphi^C_\alpha \right) - \varepsilon_{\beta\alpha} \left(S^{BA} - S_{BA} \right)$$

$$- \frac{1}{4} \varepsilon_{\beta\alpha} \left(\varphi^{\gamma B} \varphi^A_\gamma - \varphi^\gamma_B \varphi_{\gamma A} \right)$$

$$+ \varepsilon_{\beta\alpha} \delta^{BA} \left(2i \, e^{-\varphi} G - \frac{3}{4} \varphi^{\gamma C} \varphi^C_\gamma \right), \tag{15.50}$$

where

$$G^{BA} = \varepsilon^{BA} G . \tag{15.51}$$

($\mathcal{D}$ is still the full $SL(2, \mathbf{C}) \times U(2)$ covariant derivative.) Furthermore, (14.24) gives

$$\mathcal{D}_{\dot\alpha B} \varphi^A_\alpha = 4i \, \varepsilon^{BA} \, e^{-\varphi} \, \tilde{G}_{\alpha\dot\alpha} + 2 \varphi^{BA}_{\alpha\dot\alpha} + 2 \delta^A_B \, U_{\alpha\dot\alpha}$$

$$- i \delta^A_B \, \mathcal{D}_{\alpha\dot\alpha} \varphi + 2 \varphi_{\alpha B} \, \overline{\varphi}^A_{\dot\alpha} - \varphi_{\alpha A} \, \overline{\varphi}^B_{\dot\alpha} , \tag{15.52}$$

where

$$\tilde{\Phi}_{\alpha\dot\alpha}{}^B{}_A = \varepsilon^{BA} \, \tilde{\Phi}_{\alpha\dot\alpha} + i \varphi^{(BA)}_{\alpha\dot\alpha} , \qquad \varphi^{AA}_{\alpha\dot\alpha} = 0 . \tag{15.53}$$

Next we consider the SO(2) curvature, which is given by

$$\tilde{R}_{\mathcal{DC}} = \frac{1}{2} \tilde{R}_{\mathcal{DC}}{}^{AA} + \tilde{\Phi}_{\mathcal{D}}{}^A{}_B \, \tilde{\Phi}_{\mathcal{C}}{}^{BA} . \tag{15.54}$$

Its lowest-dimensional components are

$$\widetilde{R}_{\beta\,\alpha}^{\,BA} = -\delta^{BA}\left(2\,V_{\beta\alpha} - \frac{1}{2}\,\varphi_{\beta C}\,\varphi_{\alpha}^{\,C}\right)$$

$$- \varepsilon_{\beta\alpha}\,\varepsilon^{BA}\left(S^{CC} - \frac{1}{4}\,\varphi^{\gamma C}\,\varphi_{\gamma}^{\,C}\right), \tag{15.55}$$

$$\widetilde{R}_{\beta\,\dot{\alpha}A}^{\,B} = -\varepsilon^{BA}\left(2\,U_{\beta\dot{\alpha}} + \frac{1}{2}\,\varphi_{\beta}^{\,C}\,\overline{\varphi}_{\dot{\alpha}C}\right) - \frac{1}{2}\,\delta_{A}^{B}\,\varphi_{\beta}^{\,C}\,\overline{\varphi}_{\dot{\alpha}}^{\,C}. \tag{15.56}$$

In order to be able to apply the action formula (15.45), we have to bring these equations into the form of the Yang-Mills constraints (13.21). This is achieved by the redefinition

$$\widehat{\Phi}_{\alpha}^{\,A} = \widetilde{\Phi}_{\alpha}^{\,A} - \frac{1}{2}\,\varphi_{\alpha A},$$

$$\widehat{\Phi}_{\alpha\dot{\alpha}} = \widetilde{\Phi}_{\alpha\dot{\alpha}} - 2\,e^{-\varphi}\,\widetilde{G}_{\alpha\dot{\alpha}} - \frac{i}{2}\,\varphi_{\alpha}^{\,A}\,\overline{\varphi}_{\dot{\alpha}}^{\,A}, \tag{15.57}$$

which implies

$$\widehat{R}_{\beta\,\alpha}^{\,BA} = \varepsilon_{\beta\alpha}\,\varepsilon^{BA}\left(-2i\,e^{-\varphi}\,G - S^{CC} + \frac{1}{2}\,\varphi^{\gamma C}\,\varphi_{\gamma}^{\,C}\right), \tag{15.58}$$

$$\widehat{R}_{\beta\,\dot{\alpha}A}^{\,B} = 0. \tag{15.59}$$

Thus we have

$$\mathcal{S} = \int d^4x \; d^4\Theta \; \mathcal{E} \; \overline{B}\,\overline{F} + \text{c.c.}, \tag{15.60}$$

where

$$\overline{F} = 2\,e^{-\varphi}\,\overline{G} - i\,S^{AA} - \frac{i}{2}\,\overline{\varphi}_{\dot{\alpha}}^{\,A}\,\overline{\varphi}^{\dot{\alpha}A}. \tag{15.61}$$

Neither the invariant (15.5) nor the invariant (15.60) alone give consistent field equations for the multiplet defined by the constraint (15.46). Their sum, however, is a consistent action for Poincaré supergravity:

$$\mathcal{S} = \int d^4x \; d^4\Theta \; \mathcal{E} \; (1 + \overline{B}\,\overline{F}) + \text{c.c.}. \tag{15.62}$$

(A relative factor may be absorbed by a rescaling of $\overline{B}$.) It describes the *new minimal* $40 + 40$ *multiplet* [41]

$$\left(e_m{}^a,\; \psi_{mA}{}^{\alpha},\; a_m \;\middle|\; v_m,\; b_{mn},\; \varphi,\; \varphi_{\alpha}^{\,A},\; V_{\beta\alpha},\; S^{BA},\; U_{\alpha\dot{\alpha}},\; G,\; \varphi_{\alpha\dot{\alpha}}^{\,BA},\; \overline{\Lambda}_{\dot{\alpha}A},\; P\right)$$

$$(\theta = \overline{\theta} = 0). \tag{15.63}$$

New Minimal 32 + 32 Multiplet. Like the minimal 40 + 40 multiplet, the above set of fields can also be further reduced by an on-shell condition for one of the compensating multiplets. This time the victim is the vector multiplet and the additional constraint reads

$$S^{BA} = 0.$$ (15.64)

The consequences of this constraint are

$$\overline{\Lambda}_{\dot\alpha A} = -2\,\overline{\Psi}_{\dot\alpha A}\,,$$ (15.65)

$$P = \frac{1}{6}\,R + \frac{1}{3}\left(V^{\alpha\beta}\,W_{\alpha\beta} + \overline{V}_{\dot\alpha\dot\beta}\,\overline{W}^{\dot\alpha\dot\beta}\right) + \frac{1}{2}\,U^{\alpha\dot\alpha}\,U_{\alpha\dot\alpha}\,,$$ (15.66)

$$\mathcal{D}^{\beta}_{\ \alpha}\left(W_{\beta\alpha} + V_{\beta\alpha}\right) - \mathcal{D}_{\alpha}^{\ \dot\beta}\left(\overline{W}_{\dot\beta\dot\alpha} + \overline{V}_{\dot\beta\dot\alpha}\right) = 0\,,$$ (15.67)

$$\mathcal{D}^{\alpha\dot\alpha}\,U_{\alpha\dot\alpha} = i\left(V^{\alpha\beta}\,W_{\alpha\beta} - \overline{V}_{\dot\alpha\dot\beta}\,\overline{W}^{\dot\alpha\dot\beta}\right),$$ (15.68)

where $\mathcal{D}_{\alpha\dot\alpha}$ is the Lorentz covariant derivative. Again, we find that the field equations are either absorbed by fields of the supergravity multiplet or can be interpreted as Bianchi identities.

Equation (15.67) is the Bianchi identity for an additional one-form gauge potential $\widehat{A}$ describing an abelian vector field $\hat{a}_m$. The field strength

$$\widehat{F} = d\widehat{A}$$

satisfies the constraints

$$\widehat{F}^{\ BA}_{\beta\ \alpha} = i\,\varepsilon_{\beta\alpha}\,\varepsilon^{BA}\,,$$

$$\widehat{F}^{\ B}_{\beta\ \dot\alpha A} = 0.$$ (15.69)

The Bianchi identity (15.68) can be solved by a tensor field $\hat{b}_{mn}$ or, in superspace, a two-form gauge potential $\widehat{B}$. However, the non-linear terms on the right-hand side of (15.68) show that $\widehat{B}$ cannot be an ordinary two-form potential. The transformation law (14.2) has to be modified to

$$\delta\widehat{B} = d\widehat{\Omega} + \Lambda F + \widehat{\Lambda}\widehat{F}.$$ (15.70)

The invariant field strength is then given by

$$\widehat{G} = d\widehat{B} + AF + \widehat{A}\widehat{F}$$ (15.71)

and the Bianchi identity reads

$$d\widehat{G} = FF + \widehat{F}\widehat{F}.$$ (15.72)

In addition to the usual constraints (14.13–14), $\widehat{G}$ is restricted by

$$\widehat{G}^A_B = 0,\qquad(15.73)$$

where $\widehat{G}^A_B$ is the basic superfield of an ordinary tensor multiplet.

Altogether, we obtain the *new minimal $32 + 32$ multiplet* [43]

$$\left(e_m{}^a,\ \psi_{mA}{}^\alpha,\ a_m\ \middle|\ \hat{a}_m,\ \varphi^A_\alpha,\ \varphi,\ \hat{b}_{mn}\ \middle|\ v_m,\ b_{mn},\ G,\ \varphi^{BA}_{\alpha\dot\alpha}\right)$$

$$(\theta = \bar\theta = 0).\qquad(15.74)$$

It is off-shell irreducible although it describes on-shell the coupling of $N = 2$ supergravity to a vector multiplet.

The action for the above multiplet is again given by (15.62), but the constraint (15.64) simplifies this expression considerably. Using the explicit component form of the chiral density $\mathcal{E}$ [57], one can show that (15.64) implies

$$\int \mathrm{d}^4 x\ \mathrm{d}^4 \Theta\ \mathcal{E} = 0.\qquad(15.75)$$

Thus the action (15.62) becomes

$$\mathcal{S} = \int \mathrm{d}^4 x\ \mathrm{d}^4 \Theta\ \mathcal{E}\ \overline{B}\left(2\,\mathrm{e}^{-\varphi}\,\overline{G} - \frac{\mathrm{i}}{2}\,\overline{\varphi}^A_{\dot\alpha}\,\overline{\varphi}^{\dot\alpha A}\right) + \mathrm{c.\,c.}.\qquad(15.76)$$

$*$

To summarize this chapter, we have illustrated in Figure 1 the relations between the various $N = 2$ off-shell supergravity multiplets (cf. Figure 1 of Ref. [41]).

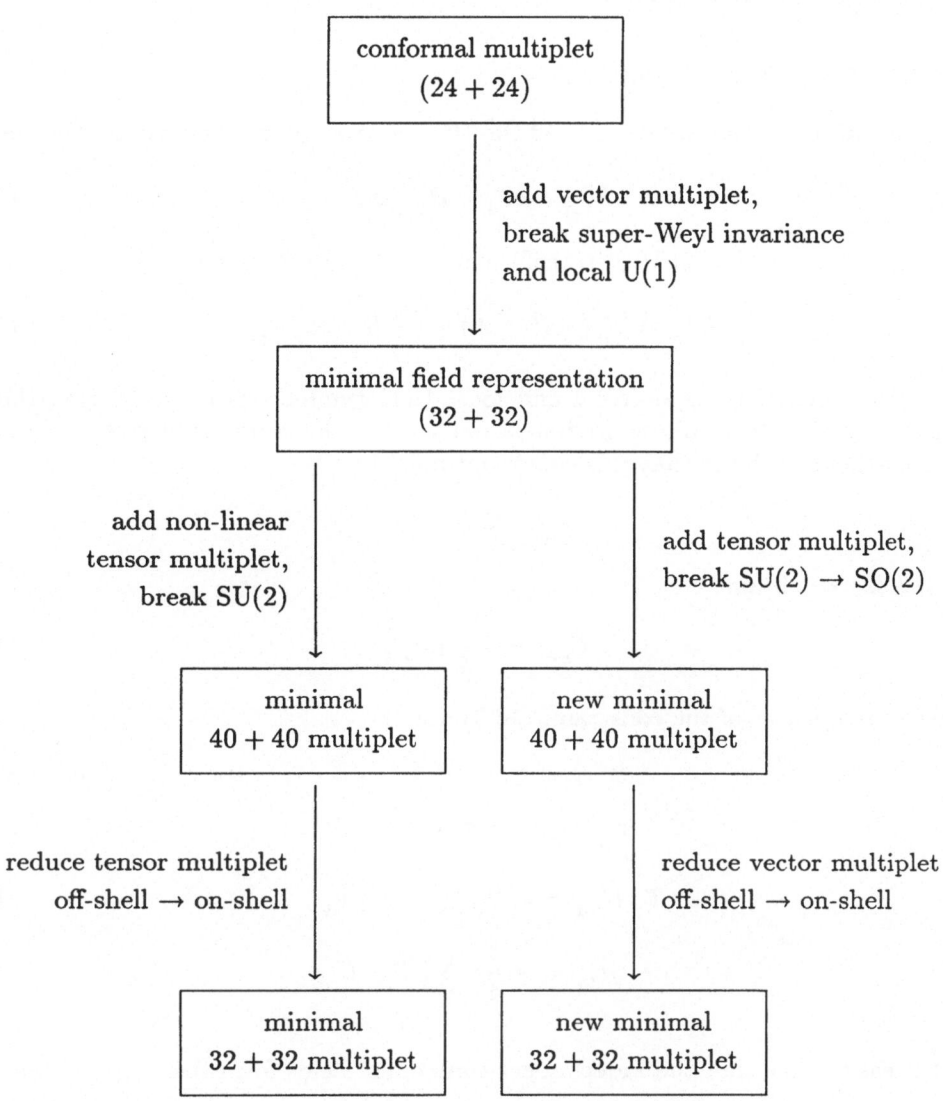

Fig. 1. $N = 2$ off shell supergravity multiplets

16. $N = 1$ Off-Shell Supergravity

We still have to discuss $N = 1$ off-shell Poincaré supergravity. Here one may again distinguish two cases: with or without local U(1) symmetry. We start with the case where the U(1) invariance is broken.

16.1 The Minimal Multiplet

The general transformation law (3.2) of the U(1) connection reads in component form

$$\delta \Phi_A = \Omega_A - H_A{}^B \Phi_B .$$ (16.1)

Under super-Weyl and local U(1) transformations, Φ_α transforms as

$$\delta \Phi_\alpha = \mathcal{D}_\alpha \left(\tfrac{3}{2} H - \Lambda \right) + \left(\tfrac{1}{2} H - \Lambda \right) \Phi_\alpha .$$ (16.2)

Now we break the super-Weyl and local U(1) symmetries by restricting the parameters H and Λ to a minimal off-shell multiplet – the chiral multiplet. The above transformation law shows that this constraint must be

$$\Phi_{\underline{\alpha}} = 0 .$$ (16.3)

From (12.32) we obtain then

$$\Phi_{\alpha\dot\alpha} = -\frac{3}{2} i \, U_{\alpha\dot\alpha} .$$ (16.4)

Further consequences of the constraint (16.3) are

$$\mathcal{D}_\alpha S = 0 ,$$

$$\mathcal{D}_{\dot\alpha} S = -4 \, \overline{\Psi}_{\dot\alpha} ,$$ (16.5)

$$\mathcal{D}_\beta U_{\alpha\dot\alpha} = -\overline{\Psi}_{\beta\alpha\dot\alpha} - \varepsilon_{\beta\alpha} \overline{\Psi}_{\dot\alpha} ,$$ (16.6)

$$\rho_{\beta\alpha} = -\frac{3}{4} i \sum_{\beta\alpha} \mathcal{D}_\beta{}^{\dot\alpha} U_{\alpha\dot\alpha} .$$ (16.7)

It is easy to see that the independent component fields form the *minimal multiplet* [34]

$$\left(e_m{}^a , \, \psi_m{}^\alpha \, | \, S , \, U_{\alpha\dot\alpha} \right) \qquad (\theta = \overline{\theta} = 0)$$ (16.8)

with 12 bosonic and 12 fermionic degrees of freedom. The superfield action for this multiplet is simply the volume of $N = 1$ superspace [20]:

$$S = \int dz \, E .$$ (16.9)

Because of $\overline{\Delta}\, 1 = 4\,\overline{S}$, the corresponding chiral action reads

$$S = \int d^4x\, d^2\Theta\, \mathcal{E}\, \overline{S} + \text{c.c.}. \tag{16.10}$$

16.2 The New Minimal Multiplet

The second possibility is to leave the local U(1) symmetry unbroken. As we explained in Section 15.3, this can only be done in the presence of a tensor multiplet which compensates for the U(1) gauge transformations. Using the super-Weyl invariance, one can impose the gauge condition

$$G = 1 \tag{16.11}$$

on the basic superfield of the tensor multiplet. The consequences of this constraint are

$$G_\alpha = 0, \tag{16.12}$$

$$S = 0, \tag{16.13}$$

$$U_{\alpha\dot\alpha} = 2\,\widetilde{G}_{\alpha\dot\alpha}. \tag{16.14}$$

Thus we obtain the *new minimal multiplet* [36]

$$\left(e_m{}^a,\ \psi_m{}^\alpha \mid v_m,\ b_{mn}\right) \qquad (\theta = \overline{\theta} = 0) \tag{16.15}$$

with $12 + 12$ degrees of freedom.

Before we construct an action for this multiplet, we note that the chiral projection operator (9.6) becomes

$$\overline{\Delta} = \mathcal{D}_{\dot\alpha}\, \mathcal{D}^{\dot\alpha} \tag{16.16}$$

because of (16.13). This implies

$$\int dz\, E\,\Phi = 0 \tag{16.17}$$

for any chiral superfield Φ. In particular, the volume of $N = 1$ superspace vanishes [58]:

$$\int dz\, E = 0. \tag{16.18}$$

Therefore the action (16.9) is not suitable for the new minimal multiplet.

A better action is the $N = 1$ analogue of (15.45),

$$S = \int d^4x\, d^2\Theta\, \mathcal{E}\, B^\alpha F_\alpha + \text{c.c.}, \tag{16.19}$$

where B_α is the chiral prepotential of the tensor multiplet containing b_{mn}. The superfield

F_α can be read off from the component $R^{\dot\beta}{}_a$ of the U(1) curvature. Using the solution of the Bianchi identities, we find

$$R_{\beta\alpha} = 0, \tag{16.20}$$

$$R_{\beta\dot\alpha} = 3\,U_{\beta\dot\alpha}, \tag{16.21}$$

$$R_{\beta\,\alpha\dot\alpha} = \frac{3}{2}\mathrm{i}\,\big(\,\overline\Psi_{\beta\alpha\dot\alpha} + \varepsilon_{\beta\alpha}\,\overline\Psi_{\dot\alpha}\,\big). \tag{16.22}$$

In order to get the Yang-Mills constraints $R_{\underline\beta\,\underline\alpha} = 0$, we define

$$\widehat\Phi_{\alpha\dot\alpha} = \Phi_{\alpha\dot\alpha} + \frac{3}{2}\mathrm{i}\,U_{\alpha\dot\alpha}, \tag{16.23}$$

which gives

$$\widehat R_{\underline\beta\,\underline\alpha} = 0, \tag{16.24}$$

$$\widehat R_{\beta\,\alpha\dot\alpha} = 6\mathrm{i}\,\varepsilon_{\beta\alpha}\,\overline\Psi_{\dot\alpha}. \tag{16.25}$$

Hence we have $F_\alpha \sim \Psi_\alpha$ and the action for the new minimal multiplet reads

$$S = \int \mathrm{d}^4 x \; \mathrm{d}^2\Theta \; \mathcal{E}\, B^\alpha \Psi_\alpha + \mathrm{c.\,c.}. \tag{16.26}$$

*

The results of this chapter are summarized in Figure 2, which shows the two $N = 1$ off-shell Poincaré multiplets and their relations to the conformal multiplet.

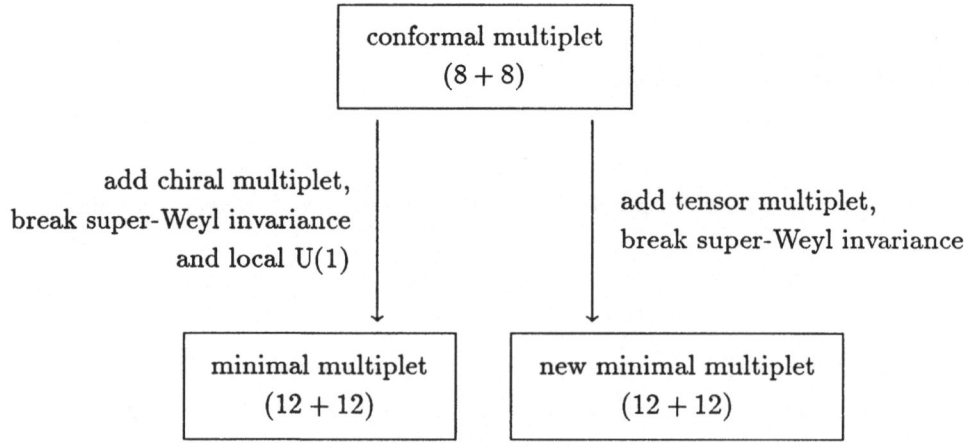

Fig. 2. $N = 1$ off-shell supergravity multiplets

Part III

Conclusion

So far we have discussed pure, classical supergravity theories. Of course, this is not yet the end of the story, and the last part of these notes will be devoted to the possible extensions. First we shall consider various couplings to matter multiplets and in the last chapter we shall give a brief outlook into quantum supergravity. In contrast to Parts I and II, the discussion will be more qualitative than quantitative, but the missing details may all be found in the cited literature.

17. Matter Couplings

The matter multiplets that may be coupled to the pure supergravity theories are the scalar multiplets (spin $\leq \frac{1}{2}$) and the vector multiplets (spin ≤ 1). Scalar multiplets exist for $N \leq 2$ and vector multiplets for $N \leq 4$. For $N > 4$, the only possible extensions are generalizations of the cosmological constant. The resulting theories are called gauged Poincaré supergravities.

While the matter couplings for $N > 2$ are severely restricted, $N \leq 2$ supergravity allows rather general couplings (with any number of derivatives). Therefore one has to reduce the number of possibilities for $N \leq 2$ by a (more or less arbitrary) assumption. In order to include the truncations of the $N > 2$ Poincaré supergravities, we choose the following one:

(D) The terms with the most derivatives are the standard kinetic (or topological) terms multiplied by scalar fields.

For $N = 1$, similar assumptions were made in Ref. [59]. Although the consequences of (D) agree with the results of [59], our assumption is actually stronger (see Ref. [60] for examples).

17.1 $N = 1$

Multiplets. The two $N = 1$ on-shell matter multiplets are the scalar multiplet (A, χ_α) and the vector multiplet (a_m, λ_α). They have each 2 bosonic and 2 fermionic degrees of freedom.

Off-shell, the scalar multiplet may be described by a chiral multiplet [61], a tensor multiplet [62] or a non-minimal multiplet [33]. The latter one is reducible to either of the former multiplets [2, first ref.]. The field content of the off-shell vector multiplet [61] is essentially unique. For convenience we have listed the minimal $(4 + 4)$ matter multiplets once more in Table 8.

multiplet	chiral	tensor	vector
gauge potential	(3-form)	2-form	1-form
basic superfield	Φ	G	F_α
constraints	$\mathcal{D}_{\dot\alpha} \Phi = 0$	$G = \overline{G},$ $\overline{\Delta} G = 0$	$\mathcal{D}_{\dot\alpha} F_\alpha = 0,$ $\mathcal{D}^\alpha F_\alpha = \mathcal{D}_{\dot\alpha} \overline{F}^{\dot\alpha}$
component fields	$(A, \chi_\alpha \mid F)$	$(b_{mn}, G, \varphi_\alpha)$	$(a_m, \lambda_\alpha \mid D)$

Table 8. $N = 1$ off-shell matter multiplets

Note that all of them can be described by p-form gauge potentials in superspace [63].

Free Actions. Before we consider general interactions of the above multiplets, we write down the free actions in a superconformal background. First of all, we remark that a chiral superfield Φ with Weyl weight ω must have U(1) weight $\frac{2}{3}\omega$ because of $\mathcal{D}_{\dot\alpha} \Phi = 0$. The free action for the chiral multiplet requires $\omega = 1$ and reads

$$S = \int dz \, E \, \Phi \overline{\Phi}, \qquad (17.1)$$

which is equivalent to

$$S = \int d^4x \, d^2\Theta \, \mathcal{E} \, \overline{\Delta} (\Phi \overline{\Phi}) + \text{c.c.}. \qquad (17.2)$$

The vector multiplet is described by a chiral superfield F_α with Weyl weight $\frac{3}{2}$ and U(1) weight 1. The action is

$$S = \int d^4x \, d^2\Theta \, \mathcal{E} \, \text{tr}(F^\alpha F_\alpha) + \text{c.c.}. \qquad (17.3)$$

The basic superfield G of the tensor multiplet has Weyl weight 2 and U(1) weight 0. The superconformal (improved) action reads [64]

$$S = \int d^4x \, d^2\Theta \, \mathcal{E} \, B^\alpha F_\alpha + \text{c.c.}, \tag{17.4}$$

where

$$F_\alpha = \overline{\Delta} \, \mathcal{D}_\alpha V \tag{17.5}$$

and

$$V = \log \frac{G}{\Phi \overline{\Phi}}. \tag{17.6}$$

B_α is the chiral prepotential of the tensor multiplet and Φ is a chiral superfield with Weyl weight 1. The only purpose of Φ is to compensate for the Weyl weight of G. The action (17.4) does not depend on Φ since (17.5) is invariant under

$$V' = V + \Lambda + \overline{\Lambda}$$

where Λ is chiral.

General Interactions. We now turn to the most general supergravity-matter couplings satisfying assumption (D). The scalar multiplets are described by (a priori unconstrained) complex superfields Φ^i and the vector multiplets by (constrained) chiral superfields F_α^I. Here the index i labels the matter representation and the index I the adjoint representation of the gauge group $\mathcal{G}$.

We define

$$\mathcal{D}_\alpha \Phi^i = \chi_\alpha^i, \tag{17.7}$$

where $\mathcal{D}$ is the full gauge covariant derivative. The $\theta = \overline{\theta} = 0$ components of χ_α^i are the physical fermions of the scalar multiplets (up to redefinitions). Therefore we may impose a constraint on $\mathcal{D}_{\dot\alpha} \Phi^i$. The most general one is

$$\mathcal{D}_{\dot\alpha} \Phi^i = f^{i\overline{k}}(\Phi, \overline{\Phi}) \, \overline{\chi}_{\dot\alpha k} + g_I^i(\Phi, \overline{\Phi}) \, \overline{F}_{\dot\alpha}^I. \tag{17.8}$$

The consistency condition for this constraint is $\{\mathcal{D}_{\dot\alpha}, \mathcal{D}_{\dot\beta}\} \Phi^i = 0$, which implies[1]

$$g_I^i = 0, \tag{17.9}$$

$$D^{\overline{k}} f^{i\overline{l}} - D^{\overline{l}} f^{i\overline{k}} = 0, \tag{17.10}$$

where

$$D^{\overline{k}} = \frac{\partial}{\partial \overline{\Phi}_k} + f^{i\overline{k}} \frac{\partial}{\partial \Phi^i}. \tag{17.11}$$

---

[1]   In addition, the Ricci identity $\{\mathcal{D}_\alpha, \mathcal{D}_{\dot\alpha}\} \Phi^i$ requires that the matrix $M_j^i = \delta_j^i - f^{i\overline{k}} \overline{f}_{\overline{k}j}$ is invertible. (I thank Julius Wess for this observation.)

Next we perform a redefinition

$$\widehat{\Phi}^i = \widehat{\Phi}^i(\Phi, \overline{\Phi}),$$

where $\widehat{\Phi}^i$ are the solutions of

$$D^{\bar{k}} \widehat{\Phi}^i = 0. \tag{17.12}$$

This is an overdetermined, but consistent system of differential equations and the solutions satisfy

$$\mathcal{D}_{\dot{\alpha}} \widehat{\Phi}^i = 0. \tag{17.13}$$

The net result of this paragraph is that $N = 1$ scalar multiplets can be described by chiral superfields without loss of generality.

Now we include $N = 1$ Poincaré supergravity and consider first the case where all the additional conformal symmetries are broken. In the presence of chiral and vector multiplets, the most general constraint that breaks the local U(1) symmetry is given by

$$V = f(\Phi, \overline{\Phi}), \tag{17.14}$$

where V is the U(1) prepotential. The function f, however, can always be absorbed by a super-Weyl transformation. Therefore it suffices to start with

$$V = 0, \tag{17.15}$$

which defines the minimal supergravity multiplet (16.8). (Later we shall come back to the constraint (17.14).)

The most general coupling of the minimal multiplet to chiral and vector multiplets is given by [65–67, 59]

$$S = \int \mathrm{d}^4 x \, \mathrm{d}^2 \Theta \, \mathcal{E} \left[\overline{\Delta} \, e^{-\frac{1}{3} K(\Phi, \overline{\Phi})} + W(\Phi) \right.$$
$$\left. + h_{(IJ)}(\Phi) \, F^{\alpha I} F_\alpha^J \right] + \text{c.c.}. \tag{17.16}$$

The Kähler potential K is a real function of Φ^i and $\overline{\Phi}_i$. The superpotential W and the functions h_{IJ} depend only on Φ^i and are therefore chiral. In order to make the symmetries of the action (17.16) manifest, we write down the corresponding superconformal action [68]:

$$S = \int \mathrm{d}^4 x \, \mathrm{d}^2 \Theta \, \mathcal{E} \left[\overline{\Delta} \, (e^{-\frac{1}{3} K(\Phi, \overline{\Phi})} \, \Psi \overline{\Psi}) + W(\Phi) \, \Psi^3 \right.$$
$$\left. + h_{IJ}(\Phi) \, F^{\alpha I} F_\alpha^J \right] + \text{c.c.}. \tag{17.17}$$

Ψ is the chiral compensator for the minimal multiplet with Weyl weight 1 and U(1) weight $\frac{2}{3}$. With the help of Ψ the chiral superfields Φ^i can be redefined such that they have vanishing Weyl and U(1) weights.

The action (17.17) is invariant under

$$K' = K + F(\Phi) + \overline{F}(\overline{\Phi}),$$

$$W' = e^{-F} W,$$

$$\Psi' = e^{\frac{1}{3}F} \Psi. \tag{17.18}$$

Thus the superfields Φ^i may be considered as the coordinates of a Kähler manifold with Kähler potential K [69].

The Kähler invariance (17.18) can be used to eliminate the superpotential W from the action. Setting

$$F = \log W, \tag{17.19}$$

one obtains

$$S = \int \mathrm{d}^4x \, \mathrm{d}^2\Theta \, \mathcal{E} \Big[\overline{\Delta} \big(e^{-\frac{1}{3}G(\Phi,\overline{\Phi})} \, \Psi_0 \, \overline{\Psi}_0 \big) + (\Psi_0)^3$$

$$+ h_{IJ}(\Phi) \, F^{\alpha I} F_\alpha^J \Big] + \text{c.c.}. \tag{17.20}$$

Finally, we fix a super-Weyl and U(1) gauge by imposing a condition on the compensating superfield Ψ_0. An obvious choice would be $\Psi_0 = \text{const}$, but this leads to a lagrangian in which the curvature scalar is multiplied by scalar fields. Therefore we choose

$$\Psi_0 = c \, e^{\frac{1}{6}G}, \tag{17.21}$$

which implies

$$\Phi_\alpha = \frac{1}{4} \, \mathcal{D}_\alpha G \tag{17.22}$$

for the U(1) connection. The U(1) prepotential is thus proportional to the function $G(\Phi,\overline{\Phi})$. Up to an overall factor, the action (17.20) becomes [70]

$$S = \int \mathrm{d}^4x \, \mathrm{d}^2\Theta \, \mathcal{E} \Big[c\overline{S} + e^{\frac{1}{2}G(\Phi,\overline{\Phi})} + h_{IJ}(\Phi) \, F^{\alpha I} F_\alpha^J \Big] + \text{c.c.}. \tag{17.23}$$

The constant c is fixed by the normalization of the curvature scalar.

Gauged $N = 1$ Supergravity. In Section 16.2 we have seen that it is possible to leave the local U(1) symmetry unbroken. However, the U(1) gauge field was only an auxiliary field. In the following we shall consider the case that the gauge field survives on-shell, i.e., the action contains a kinetic term for the U(1). This is called $N = 1$ Poincaré supergravity with a gauged U(1) symmetry (R-invariance) [71, 35, 72].

The most general supergravity-matter coupling with a gauged U(1) is given by the action (17.16), where one of the superfields F_α^I describes the U(1) gauge field. However, the functions K, W, and h_{IJ} are now restricted by the U(1) invariance [73]. That is, the U(1) weights $w(\Phi^i)$ must be assigned such that $w(K) = w(h_{IJ}) = 0$ and $w(W) = 2$. The corresponding superconformal action is

$$S = \int d^4x \, d^2\Theta \, \mathcal{E} \left[\overline{\Delta} \left(e^{-\frac{1}{3}K(\Phi,\overline{\Phi})} \Psi \overline{\Psi} \right) + W(\Phi) \right.$$
$$\left. + h_{IJ}(\Phi) \, F^{\alpha I} F_\alpha^J \right] + \text{c.c.} \qquad (17.24)$$

with the Kähler invariance

$$K' = K + F(\Phi) + \overline{F}(\overline{\Phi}),$$

$$\Psi' = e^{\frac{1}{3}F} \Psi. \qquad (17.25)$$

One of the superfields F_α^I in (17.24) is abelian and the chiral compensator Ψ has a non-vanishing weight under this abelian group (in addition to the Weyl and U(1) weights).

As in (17.21), we now impose the gauge condition

$$\Psi = c \, e^{\frac{1}{6}K}, \qquad (17.26)$$

which leads to

$$\Phi_\alpha = \frac{1}{4} \, \mathcal{D}_\alpha K + A_\alpha. \qquad (17.27)$$

Φ_α is the U(1) connection and A_α the connection of the abelian group. The action (17.24) becomes then

$$S = \int d^4x \, d^2\Theta \, \mathcal{E} \left[c \overline{S} + W(\Phi) + h_{IJ}(\Phi) \, F^{\alpha I} F_\alpha^J \right] + \text{c.c..} \qquad (17.28)$$

The superfield $\overline{S}$ contains, besides the curvature scalar, a Fayet-Iliopoulos term for the U(1) [74].

We conclude this section with two remarks. First, it is possible to gauge all the isometries of the Kähler manifold $\{\Phi^i, \overline{\Phi}_i\}$, not only those which are linearly realized [67, 75]. Second, the most general matter couplings of the new minimal multiplet (16.15) are on-shell equivalent to the U(1) invariant couplings of the minimal multiplet [76].

17.2 $N = 2$

Multiplets. The two $N = 2$ on-shell matter multiplets are the scalar multiplet $\left(A_A^i, \chi_\alpha^i\right)$ $(i = 1, 2)$ and the vector multiplet $\left(a_m, \lambda_{\alpha A}, F\right)$. They have each 4 bosonic and 4 fermionic degrees of freedom.

There are various off-shell versions of the scalar multiplet: the hypermultiplet with a central charge [77] or with an infinite number of auxiliary fields [47], the tensor multiplet [6], the relaxed hypermultiplet [78], and the first-order multiplet [79]. For the vector multiplet, there exist only two off-shell versions: the standard multiplet [77, 80] and the multiplet with a central charge [81]. The minimal $(8 + 8)$ matter multiplets without central charges are once more listed in Table 9.

multiplet	tensor	vector
gauge potential	2-form	1-form
basic superfield	G_B^A	F
constraints	$G_B^A = \overline{G}_B^A, \; G_A^A = 0,$ $\mathcal{D}_\alpha^{(A} G^{BC)} = 0$	$\mathcal{D}_\alpha^A F = 0,$ $F_B^A = \overline{F}_B^A$
component fields	$\left(b_{mn}, G_B^A, \varphi_\alpha^A \,\middle|\, H\right)$	$\left(a_m, \lambda_{\alpha A}, F \,\middle|\, D_B^A\right)$

Table 9. $N = 2$ off-shell matter multiplets

Note that both multiplets together form a chiral superfield.

Free Actions. The free action for the vector multiplet in a superconformal background reads

$$S = \int \mathrm{d}^4 x \, \mathrm{d}^4 \Theta \; \mathcal{E} \, \mathrm{tr}\left(\overline{F}\, F\right) + \text{c.c..} \tag{17.29}$$

This expression is super-Weyl invariant since the chiral superfield $\overline{F}$ has Weyl weight 1 and U(1) weight 2.

The tensor multiplet is described by a hermitian superfield G_B^A with Weyl weight 2 and U(1) weight 0. The superconformal (improved) action is [64, 82, 41]

$$S = \int \mathrm{d}^4 x \, \mathrm{d}^4 \Theta \; \mathcal{E} \, \overline{B}\, F + \text{c.c.,} \tag{17.30}$$

where

$$\overline{F} = g^{-1}\left(\overline{G} + \tfrac{1}{2} G^{AB}\, \overline{S}_{AB}\right) + 2\, g^{-3} G^{AB}\, \overline{G}_{\dot{\alpha}A}\, \overline{G}_B^{\dot{\alpha}} \tag{17.31}$$

and

$$g = \sqrt{G_B^A\, G_A^B}. \tag{17.32}$$

$\overline{B}$ is the chiral prepotential of the tensor multiplet. The action (17.30) is invariant under the gauge transformations of $\overline{B}$ because $\overline{F}$ satisfies the constraints of a vector superfield (see Table 9).

Actions for Scalar Multiplets. The most general on-shell coupling of scalar multiplets to $N = 2$ Poincaré supergravity was given in Ref. [83]. It was shown that the scalar fields can be considered as the coordinates of a quaternionic manifold. Unfortunately, these results cannot easily be extended off the mass-shell because there is no simple off-shell scalar multiplet. The only off-shell version that allows general interactions is the hypermultiplet, which needs either a central charge or an infinite number of auxiliary fields.

General actions for the hypermultiplet with a central charge were constructed in Refs. [84–86]. However, the central charge excludes an unconstrained superfield formulation and is not compatible with the full superconformal group. Both disadvantages can be avoided by using the "harmonic" hypermultiplet with an infinite number of auxiliary fields [44, 87]. It seems, therefore, that the harmonic superspace is the most suitable framework for the description of general $N = 2$ supergravity-matter couplings involving scalar multiplets.

Actions for Vector Multiplets. In the following we shall restrict ourselves to supergravity-matter couplings involving only vector multiplets. The basic superfields of the vector multiplets are Lie algebra valued, i. e.,

$$F = F^I T_I,$$

$$\overline{F} = -\overline{F}^I T_I, \tag{17.33}$$

where

$$[T_I, T_J] = f_{[IJ]}{}^K T_K. \tag{17.34}$$

T_I are the antihermitian generators and $f_{IJ}{}^K$ the real structure constants of the gauge group $\mathcal{G}$. The transformation law of $\overline{F}^I$ is

$$\delta \overline{F}^I = \overline{F}^J \Lambda^K f_{JK}{}^I. \tag{17.35}$$

The most general coupling of vector multiplets to $N = 2$ Poincaré supergravity is given by [84, 75]

$$S = \int d^4x \, d^4\Theta \, \mathcal{E} \, h(\overline{F}^I) + \text{c.c.} \tag{17.36}$$

up to a small modification that will be discussed later. The form of the action is the same for all the off-shell supergravity multiplets described in Chapter 15. The corresponding superconformal action reads [84]

$$S = \int d^4x \, d^4\Theta \, \mathcal{E} \, h(\overline{F}^I \overline{F}_0^{-1}) \, \overline{F}_0 \overline{F}_0 + \text{c.c.}, \tag{17.37}$$

where $\overline{F}_0$ is the abelian compensator for the minimal field representation (15.22). It is convenient to give up the distinction between $\overline{F}^I$ and $\overline{F}_0$, and to start from the general superconformal action [88, 85]

$$S = \int \mathrm{d}^4 x \, \mathrm{d}^4 \Theta \, \mathcal{E} \, h\big(\overline{F}^I\big) + \mathrm{c.c.}. \tag{17.38}$$

The function h must have Weyl weight 2 and $U(1)$ weight 4, i.e.,

$$\overline{F}^I \frac{\partial h}{\partial \overline{F}^I} = 2h. \tag{17.39}$$

We still have to discuss the gauge invariance of the action (17.38). Of course, invariance of h implies invariance of S. However, it was shown in Ref. [85] that this is not the most general case. In flat superspace, S remains invariant if h transforms as

$$\delta h = \mathrm{i}\, c_{I(JK)} \, \Lambda^I \, \overline{F}^J \, \overline{F}^K, \tag{17.40}$$

where c_{IJK} are real constants. In curved superspace, this is no longer true since Λ is z-dependent. Surprisingly, however, the variation (17.40) can be compensated by adding a suitable "counterterm" to the action.

Before we construct this term, we derive some identities that will be needed later on. First of all, the transformation law (17.40) can be written as

$$\frac{\partial h}{\partial \overline{F}^I} \, \overline{F}^J \Lambda^K f_{JK}{}^I = \mathrm{i}\, c_{IJK} \, \Lambda^I \, \overline{F}^J \, \overline{F}^K. \tag{17.41}$$

Replacing Λ^I by $\overline{F}^I$, one obtains

$$c_{(IJK)} = 0. \tag{17.42}$$

Next we exploit the algebra of gauge transformations

$$\left[\delta\big(\Lambda_2^I\big), \delta\big(\Lambda_1^I\big) \right] = \delta\big(\Lambda_1^J \Lambda_2^K f_{JK}{}^I\big). \tag{17.43}$$

The closure of this algebra on h implies

$$f_{IJ}{}^M c_{MKL} = - \sum_{IJ} \hat{\sum_{KL}} f_{IK}{}^M c_{JLM} \tag{17.44}$$

and the closure on $\overline{F}^I$ gives

$$\sum_{IJK} f_{IJ}{}^M f_{MK}{}^L = 0. \tag{17.45}$$

The last relation can also be obtained from the Jacobi identity for the generators T_I.

Consider now a real 4-form H with the transformation law

$$\delta H = c_{IJK}\, \Lambda^I F^J F^K, \qquad (17.46)$$

where F^I are the field strength 2-forms. Up to an invariant 4-form, H is given by

$$H = -\frac{2}{3}\, c_{IJK}\, A^I A^J \left(F^K - \frac{1}{8} f_{LM}{}^K A^L A^M \right) + \mathrm{d}C, \qquad (17.47)$$

where

$$\delta C = \frac{2}{3}\left(c_{IJK} - c_{JIK}\right) \Lambda^I A^J F^K$$
$$+ \frac{1}{3}\, f_{IJ}{}^M c_{KLM}\, A^I A^J A^K A^L + \mathrm{d}\Sigma. \qquad (17.48)$$

The Bianchi identity for H reads

$$\mathrm{d}H = - c_{IJK}\, A^I F^J F^K. \qquad (17.49)$$

Because of the transformation law (17.46), this can also be written as

$$\mathcal{D}H = 0, \qquad (17.50)$$

which is equivalent to

$$\sum_{\mathcal{EDCBA}} \left(\mathcal{D}_{\mathcal{E}}\, H_{\mathcal{DCBA}} + 2\, T_{\mathcal{ED}}{}^{\mathcal{F}} H_{\mathcal{FCBA}} \right) = 0. \qquad (17.51)$$

In the next step, we impose the constraint

$$H^{\mathcal{DCBA}}_{\delta\gamma\beta\alpha} = \frac{1}{4} \sum_{\underline{\delta\gamma\beta\alpha}} \varepsilon_{\delta\gamma}\, \varepsilon_{\beta\alpha}\, \varepsilon^{\mathcal{DC}} \varepsilon^{\mathcal{BA}}\, \mathcal{H} \qquad (17.52)$$

and restrict the other components of H such that all of them can be expressed in terms of the superfield $\mathcal{H}$ and its covariant derivatives. The only condition on $\mathcal{H}$ from the Bianchi identities (17.51) is

$$\mathcal{D}^{\dot{A}}_{\dot{\alpha}}\, \mathcal{H} = 0. \qquad (17.53)$$

That is, $\overline{\mathcal{H}}$ is a chiral superfield with the transformation law

$$\delta \overline{\mathcal{H}} = c_{IJK}\, \Lambda^I \, \overline{F}^J \, \overline{F}^K. \qquad (17.54)$$

Hence

$$\mathcal{S} = \int \mathrm{d}^4 x\; \mathrm{d}^4\Theta\; \mathcal{E} \left[h(\overline{F}^I) - \mathrm{i}\, \overline{\mathcal{H}} \right] + \text{c.c.} \qquad (17.55)$$

is a gauge and super-Weyl invariant chiral action, which describes the most general coupling of vector multiplets to $N = 2$ supergravity. The component form of this action was first given in Ref. [85].

17.3 $N > 2$

The only matter multiplets for $N > 2$ are the $N = 3$ and $N = 4$ vector multiplets. In conventional extended superspace these multiplets are on-shell, as we have seen in Chapter 13. In harmonic superspace the $N = 3$ vector multiplet can be extended off-shell [89], and an off-shell $N = 4$ vector multiplet with central charges was found in Ref. [81].

General couplings of vector multiplets to $N = 3$ Poincaré supergravity were constructed in Ref. [90] using the group-manifold approach. For $N = 4$, general supergravity-matter couplings were derived in Ref. [54] from a superconformal starting point. In the latter case, the lagrangian contains one arbitrary angle α for each factor of the gauge group (see Section 13.5).

For $N > 4$, the only possible extensions of the pure Poincaré supergravities are generalizations of the cosmological constant. Supersymmetry then implies that the global $SO(N)$ invariance (or a related symmetry) is gauged by the vector fields of the supergravity multiplet. The resulting gauged Poincaré supergravities were constructed explicitly for $N = 2, 3$ [91], $N = 4$ [92], $N = 5$ [93], and $N = 8$ [94]. In superspace, the $SO(8)$ gauging is obtained by adding products of scalar fields on the right-hand sides of Eqs. (12.27) and (12.29) [95]. The most general gaugings for $N = 8$ were given in Ref. [96].

18. Outlook into Quantum Supergravity

We conclude these notes with some remarks on quantum supergravity. More details may be found e. g. in Refs. [24,4]. The first remark is that there is at present no evidence whatsoever that gravity is quantized at all. Nonetheless it is a reasonable assumption that the basic physical laws should be the same for all kinds of matter. This is our starting point for the following considerations.

18.1 Poincaré Supergravity

Let us begin with the case $N = 0$, i. e., Einstein gravity. The lagrangian is essentially the curvature scalar divided by Newton's constant. Since this constant has mass dimension -2, the theory is not power-counting renormalizable. This is not yet fatal, but it means that the quantum theory has to be finite at all orders to save predictability.

In Ref. [97] it was shown that Einstein gravity is finite at the one-loop level. The reason is simply that all possible counterterms vanish on-shell. At two loops, however, the theory was recently found to be divergent [98]. Hence Einstein gravity makes no sense as a quantum theory.

$N \geq 1$ Poincaré supergravity is also non-renormalizable, but it was shown to be finite even at the two-loop level [99]. Beyond two loops, however, the situation is rather unclear. On the one hand, possible on-shell counterterms are known for all N [100]. On the other hand, the coefficients in front of these counterterms are not known. At best, all of them are zero – yet there seems to be no way to prove it. This is the dilemma of quantum Poincaré supergravity.

18.2 Conformal Supergravity

Let us again begin with the case $N = 0$, i. e., Weyl gravity. The lagrangian contains the square of the Weyl tensor, which is multiplied by a dimensionless constant. Therefore the theory is power-counting renormalizable [101]. However, the renormalizability has a high price: unitarity. Because of the higher derivatives, the action describes a massive spin-2 ghost in addition to the graviton [101, 102]. Hence perturbative Weyl gravity is inconsistent.

For $N \geq 1$ conformal supergravity, the situation is essentially the same. The *linearized* theories are power-counting renormalizable, but include massive ghost multiplets [103, 8]. This is the end of the story.

*   *   *

I. e., not quite. The attentive reader will have noticed that there is a weak point in the above argument: the linear approximation may not be sufficient for $N = 4$ because the multiplet contains scalar fields with dimension 0. Indeed, we have seen in Section 9.3 that the square of the Weyl tensor is multiplied by a function of these scalars. If the vacuum expectation value of this function vanishes, the higher-derivative term does not contribute to the graviton propagator. A more detailed analysis of this case will be published elsewhere.

Appendix A

Conventions

We use the conventions of the book by Wess and Bagger [1]. In addition, we define

$$V_{\mathcal{A}} = (V_a, V_\alpha^A, V_A^{\dot\alpha}),$$

$$V_{\underline{\alpha}} = (V_\alpha^A, V_A^{\dot\alpha}),$$

$$V_{\underline{a}} = (V_a, V_\alpha^A). \tag{A.1}$$

For $N = 2$, the internal indices are raised and lowered with ε^{AB} and ε_{AB} like the spinor indices.

Complex conjugation is defined by

$$V_{\mathcal{A}}^* = (\overline{V}_a, \overline{V}_{\dot\alpha A}, \overline{V}^{\alpha A}) \tag{A.2}$$

and

$$(U_{\mathcal{A}} V_{\mathcal{B}})^* = (-)^{ab} U_{\mathcal{A}}^* V_{\mathcal{B}}^*, \tag{A.3}$$

where $a = 0$ if $\mathcal{A}$ is bosonic and $a = 1$ if $\mathcal{A}$ is fermionic. In particular, this implies

$$(U^{\mathcal{A}} V_{\mathcal{A}})^* = \overline{U}^{\mathcal{A}} \overline{V}_{\mathcal{A}}. \tag{A.4}$$

The vielbein is real,

$$\overline{E}^{\mathcal{A}} = E^{\mathcal{A}}, \tag{A.5}$$

and the connection satisfies

$$\overline{\Phi}_{\mathcal{B}}{}^{\mathcal{A}} = (-)^{ba} \Phi_{\mathcal{B}}{}^{\mathcal{A}}. \tag{A.6}$$

Therefore the tensors $E_M{}^{\mathcal{A}}$, $\Phi_{M\mathcal{B}}{}^{\mathcal{A}}$, $T_{C\mathcal{B}}{}^{\mathcal{A}}$, and $R_{DC\mathcal{B}}{}^{\mathcal{A}}$ are conjugated as if they were products of real vectors. For example,

$$(T^C{}_{\dot\beta B}{}^a)^* = -T_{\dot\gamma C}{}^{Ba}_{\ \ \beta} = -T^B{}_{\beta\dot\gamma C}{}^a,$$

$$(R^{DCB}_{\delta\gamma}{}_A)^* = -R_{\dot\delta D\dot\gamma C}{}_B{}^A = R_{\dot\delta D\dot\gamma C}{}^A{}_B. \tag{A.7}$$

In the definition of a new tensor, an eventual (anti)symmetry of the indices is marked by round (square) brackets, e. g., $X_{(\alpha\beta)}$ or $X^{[AB]}$. These brackets are then omitted in the following. (If needed, they can be found in the index on p. 125.)

The symmetry operator $\sum$ is defined by

$$\sum_{AB} X_{AB} = X_{AB} - (-)^{ab} X_{BA} .$$
(A.8)

That is, $\sum$ antisymmetrizes bosonic and symmetrizes fermionic indices (without normalization factor). In particular, the graded commutator is

$$[\mathcal{D}_A , \mathcal{D}_B \} = \sum_{AB} \mathcal{D}_A \mathcal{D}_B .$$
(A.9)

Furthermore, we need the symmetrizer $\widehat{\sum}$ for internal indices:

$$\widehat{\sum_{AB}} X^{AB} = X^{AB} + X^{BA} .$$
(A.10)

The completely traceless part of a tensor is denoted by tl, e. g.,

$$\text{tl } X^A{}_B = X^A{}_B - \frac{1}{N} \delta^A_B X^C{}_C .$$
(A.11)

Appendix B

Differential Forms

Here we list some rules for differential forms in superspace [1]. A 0-form is simply a function $f(z)$. A basis of 1-forms is given by the coordinate differentials $dz^{\mathcal{M}}$. The exterior product of two differentials satisfies

$$dz^{\mathcal{M}} dz^{\mathcal{N}} = -(-)^{mn} dz^{\mathcal{N}} dz^{\mathcal{M}}, \tag{B.1}$$

where $m = 0$ if $\mathcal{M}$ is bosonic and $m = 1$ if $\mathcal{M}$ is fermionic.

A p-form Ω_p is given by

$$\Omega_p = \frac{1}{p!} dz^{\mathcal{M}_1} \ldots dz^{\mathcal{M}_p} \, \Omega_{\mathcal{M}_p \ldots \mathcal{M}_1}. \tag{B.2}$$

The components $\Omega_{\mathcal{M}_p \ldots \mathcal{M}_1}$ are graded antisymmetric in their indices. From (B.1) follows

$$\Omega_p \, \Omega_q = (-)^{pq} \, \Omega_q \, \Omega_p. \tag{B.3}$$

The exterior derivative of a p-form Ω_p is defined by

$$d\Omega_p = \frac{1}{p!} dz^{\mathcal{M}_1} \ldots dz^{\mathcal{M}_p} dz^{\mathcal{N}} \partial_{\mathcal{N}} \, \Omega_{\mathcal{M}_p \ldots \mathcal{M}_1}. \tag{B.4}$$

This definition implies

$$d\left(\Omega_p \, \Omega_q\right) = \Omega_p \, d\Omega_q + (-)^q \left(d\Omega_p\right) \Omega_q, \tag{B.5}$$

$$dd = 0. \tag{B.6}$$

Appendix C

Useful Formulas

In this appendix we collect some useful formulas of tensor and spinor algebra. Most of them are taken from Ref. [104].

SL(2,C)

The invariant tensor of SL(2,C) is $\varepsilon_{[\alpha\beta]}$. Its inverse $\varepsilon^{[\alpha\beta]}$ is defined by

$$\varepsilon^{\alpha\beta}\varepsilon_{\beta\gamma} = \delta^\alpha_\gamma \tag{C.1}$$

and the complex conjugate is

$$(\varepsilon_{\alpha\beta})^* = \varepsilon_{\dot\alpha\dot\beta}. \tag{C.2}$$

The ε-tensor raises and lowers spinor indices:

$$\Psi^\alpha = \varepsilon^{\alpha\beta}\Psi_\beta,$$

$$\Psi_\alpha = \varepsilon_{\alpha\beta}\Psi^\beta. \tag{C.3}$$

The basic identity for the ε-tensor reads

$$\varepsilon_{\alpha\beta}\varepsilon_{\gamma\delta} + \varepsilon_{\alpha\gamma}\varepsilon_{\delta\beta} + \varepsilon_{\alpha\delta}\varepsilon_{\beta\gamma} = 0. \tag{C.4}$$

The right-hand side of this equation must be zero because the left-hand side is antisymmetric in three indices. Some consequences of this identity are

$$X_{\alpha\beta} - X_{\beta\alpha} = \varepsilon_{\alpha\beta}X^\gamma{}_\gamma,$$

$$X_{\alpha\beta} = \frac{1}{2}\sum_{\alpha\beta}X_{\alpha\beta} + \frac{1}{2}\varepsilon_{\alpha\beta}X^\gamma{}_\gamma, \tag{C.5}$$

$$X_{[\alpha\beta\gamma]} = 0. \tag{C.6}$$

SU(2)

The invariant tensor of SU(2) is $\varepsilon^{[AB]}$. Its inverse $\varepsilon_{[AB]}$ is defined by

$$\varepsilon_{AB}\varepsilon^{BC} = \delta^C_A. \tag{C.7}$$

With this definition, Eqs. (C.3–6) are also valid for ε^{AB} and ε_{AB}. Observe, however, that complex conjugation gives

$$(\varepsilon^{AB})^* = -\varepsilon_{AB}. \tag{C.8}$$

SO(3,1)

The invariant tensors of SO(3,1) are $\eta_{(ab)}$ and $\varepsilon_{[abcd]}$. The inverse tensor $\eta^{(ab)}$ is defined by

$$\eta^{ab}\,\eta_{bc} = \delta_c{}^a . \tag{C.9}$$

The basic identity for the ε-tensor is

$$\varepsilon^{abcd}\,\varepsilon_{efgh} = -\sum_{abcd} \delta_e{}^a\,\delta_f{}^b\,\delta_g{}^c\,\delta_h{}^d . \tag{C.10}$$

(The minus sign is due to the signature of η.) The contractions of this equation are

$$\varepsilon^{abcd}\,\varepsilon_{efgd} = -\sum_{abc} \delta_e{}^a\,\delta_f{}^b\,\delta_g{}^c , \tag{C.11}$$

$$\varepsilon^{abcd}\,\varepsilon_{efcd} = -2\sum_{ab} \delta_e{}^a\,\delta_f{}^b , \tag{C.12}$$

$$\varepsilon^{abcd}\,\varepsilon_{ebcd} = -3!\,\delta_e{}^a , \tag{C.13}$$

$$\varepsilon^{abcd}\,\varepsilon_{abcd} = -4! . \tag{C.14}$$

Further consequences of (C.10) are the duality relations

$$X_{[ab]} = \frac{1}{2}\,\varepsilon_{abcd}\,Y^{cd} , \qquad Y^{cd} = -\frac{1}{2}\,\varepsilon^{abcd}\,X_{ab} , \tag{C.15}$$

$$X_{[abc]} = \varepsilon_{abcd}\,Y^{d} , \qquad Y^{d} = -\frac{1}{3!}\,\varepsilon^{abcd}\,X_{abc} , \tag{C.16}$$

$$X_{[abcd]} = \varepsilon_{abcd}\,Y , \qquad Y = -\frac{1}{4!}\,\varepsilon^{abcd}\,X_{abcd} , \tag{C.17}$$

$$X_{[abcde]} = 0 . \tag{C.18}$$

SU($N > 2$)

The invariant tensor of SU(N) is $\varepsilon^{[A_1\cdots A_N]}$. For $N > 2$, we define complex conjugation by

$$\left(\varepsilon^{A_1\cdots A_N}\right)^* = \varepsilon_{A_1\cdots A_N} . \tag{C.19}$$

The basic identity for the ε-tensor is

$$\varepsilon_{A_1\cdots A_N}\,\varepsilon^{B_1\cdots B_N} = \sum_{A_1\cdots A_N} \delta_{A_1}^{B_1}\cdots\delta_{A_N}^{B_N} , \tag{C.20}$$

which implies

$$\varepsilon_{A_1 \ldots A_N} \, \varepsilon^{B_1 \ldots B_k A_{k+1} \ldots A_N} = (N-k)! \sum_{A_1 \ldots A_k} \delta^{B_1}_{A_1} \ldots \delta^{B_k}_{A_k} \qquad (C.21)$$

and

$$X^{[A_1 \ldots A_k]} = \frac{1}{(N-k)!} \, \varepsilon^{A_1 \ldots A_N} \, Y_{A_{k+1} \ldots A_N} \, ,$$

$$Y_{A_{k+1} \ldots A_N} = \frac{1}{k!} \, \varepsilon_{A_1 \ldots A_N} \, X^{A_1 \ldots A_k} \, , \qquad (C.22)$$

$$X^{[A_1 \ldots A_{N+1}]} = 0 \, . \qquad (C.23)$$

For $N = 8$, we have imposed in Chapter 11 a self-duality constraint of the form

$$X^{[ABCD]} = \frac{1}{4!} \, \varepsilon^{ABCDEFGH} \, \overline{X}_{EFGH} \, . \qquad (C.24)$$

This constraint yields the following restrictions on the product of X and $\overline{X}$ [105]:

$$X^{ABCG} \, \overline{X}_{DEFG} = \frac{1}{3! \, 4!} \sum_{ABC} \sum_{DEF} \left(9 \, \delta^A_D \, X^{BCGH} \, \overline{X}_{EFGH} \right.$$

$$\left. - \frac{1}{4} \, \delta^A_D \, \delta^B_E \, \delta^C_F \, X^{GHKL} \, \overline{X}_{GHKL} \right) , \qquad (C.25)$$

$$X^{ACDE} \, \overline{X}_{BCDE} = \frac{1}{8} \, \delta^A_B \, X^{CDEF} \, \overline{X}_{CDEF} \, . \qquad (C.26)$$

SL(2,C) ↔ SO(3,1)

Any Lorentz vector V_a can be transformed into a bispinor $V_{\alpha\dot\alpha}$ and vice versa:

$$V_{\alpha\dot\alpha} = \sigma^a_{\alpha\dot\alpha} \, V_a \, ,$$

$$V_a = -\frac{1}{2} \, \overline{\sigma}^{\dot\alpha\alpha}_a \, V_{\alpha\dot\alpha} \qquad \left(\overline{\sigma}^{\dot\alpha\alpha}_a = \sigma^{\alpha\dot\alpha}_a \right) . \qquad (C.27)$$

The σ-matrices are hermitian,

$$\left(\sigma^a_{\alpha\dot\alpha} \right)^\dagger = \sigma^a_{\alpha\dot\alpha} \, , \qquad (C.28)$$

and satisfy the basic identities

$$\sigma^a_{\alpha\dot\alpha} \, \sigma^b_{\beta\dot\beta} \, \eta_{ab} = \eta_{\alpha\dot\alpha\,\beta\dot\beta} = -2 \, \varepsilon_{\alpha\beta} \, \varepsilon_{\dot\alpha\dot\beta} \, , \qquad (C.29)$$

$$\sigma^a_{\alpha\dot\alpha} \, \sigma^b_{\beta\dot\beta} \, \sigma^c_{\gamma\dot\gamma} \, \sigma^d_{\delta\dot\delta} \, \varepsilon_{abcd} = \varepsilon_{\alpha\dot\alpha\,\beta\dot\beta\,\gamma\dot\gamma\,\delta\dot\delta}$$

$$= 4\mathrm{i} \left(\varepsilon_{\alpha\beta} \, \varepsilon_{\gamma\delta} \, \varepsilon_{\dot\alpha\dot\gamma} \, \varepsilon_{\dot\beta\dot\delta} - \varepsilon_{\alpha\gamma} \, \varepsilon_{\beta\delta} \, \varepsilon_{\dot\alpha\dot\beta} \, \varepsilon_{\dot\gamma\dot\delta} \right) . \qquad (C.30)$$

From the first identity one obtains

$$\operatorname{tr}\left(\sigma^{a}\,\overline{\sigma}^{b}\right) = \operatorname{tr}\left(\overline{\sigma}^{a}\sigma^{b}\right) = -2\,\eta^{ab}, \tag{C.31}$$

$$\left(\sigma^{a}\,\overline{\sigma}^{b} + \sigma^{b}\,\overline{\sigma}^{a}\right)_{\alpha}{}^{\beta} = -2\,\eta^{ab}\,\delta_{\alpha}^{\beta},$$

$$\left(\overline{\sigma}^{a}\sigma^{b} + \overline{\sigma}^{b}\sigma^{a}\right)^{\dot{\alpha}}{}_{\dot{\beta}} = -2\,\eta^{ab}\,\delta_{\dot{\beta}}^{\dot{\alpha}}. \tag{C.32}$$

The matrices $\sigma^{[ab]}_{(\alpha\beta)}$ and $\overline{\sigma}^{[ab]}_{(\dot{\alpha}\dot{\beta})}$ are defined by

$$\left(\sigma^{ab}\right)_{\alpha}{}^{\beta} = \frac{1}{4}\left(\sigma^{a}\,\overline{\sigma}^{b} - \sigma^{b}\,\overline{\sigma}^{a}\right)_{\alpha}{}^{\beta},$$

$$\left(\overline{\sigma}^{ab}\right)^{\dot{\alpha}}{}_{\dot{\beta}} = \frac{1}{4}\left(\overline{\sigma}^{a}\sigma^{b} - \overline{\sigma}^{b}\sigma^{a}\right)^{\dot{\alpha}}{}_{\dot{\beta}}. \tag{C.33}$$

They satisfy

$$\left(\sigma^{ab}{}_{\alpha}{}^{\beta}\right)^{\dagger} = -\,\overline{\sigma}^{ab\,\dot{\beta}}{}_{\dot{\alpha}} \tag{C.34}$$

and

$$\sigma^{a}\,\overline{\sigma}^{b} = -\eta^{ab} + 2\,\sigma^{ab},$$

$$\overline{\sigma}^{a}\sigma^{b} = -\eta^{ab} + 2\,\overline{\sigma}^{ab}, \tag{C.35}$$

$$\sum_{ab}\sigma^{a}_{\alpha\dot{\alpha}}\,\sigma^{b}_{\beta\dot{\beta}} = -2\,\varepsilon_{\dot{\alpha}\dot{\beta}}\,\sigma^{ab}_{\alpha\beta} + 2\,\varepsilon_{\alpha\beta}\,\overline{\sigma}^{ab}_{\dot{\alpha}\dot{\beta}}, \tag{C.36}$$

$$\varepsilon_{abcd}\,\sigma^{cd} = -2\mathrm{i}\,\sigma_{ab},$$

$$\varepsilon_{abcd}\,\overline{\sigma}^{cd} = 2\mathrm{i}\,\overline{\sigma}_{ab}. \tag{C.37}$$

Some other useful identities for products of σ-matrices are listed below.

$$\sigma^{ab}_{\alpha\beta}\,\sigma_{b\,\gamma\dot{\gamma}} = -\frac{1}{2}\sum_{\alpha\beta}\varepsilon_{\alpha\gamma}\,\sigma^{a}_{\beta\dot{\gamma}},$$

$$\overline{\sigma}^{ab}_{\dot{\alpha}\dot{\beta}}\,\sigma_{b\,\gamma\dot{\gamma}} = \frac{1}{2}\sum_{\dot{\alpha}\dot{\beta}}\varepsilon_{\dot{\alpha}\dot{\gamma}}\,\sigma^{a}_{\gamma\dot{\beta}}, \tag{C.38}$$

$$\sigma^{ab}\sigma^{c} = \frac{1}{2}\sum_{ab}\eta^{ac}\sigma^{b} + \frac{\mathrm{i}}{2}\,\varepsilon^{abcd}\,\sigma_{d},$$

$$\overline{\sigma}^{ab}\,\overline{\sigma}^{c} = \frac{1}{2}\sum_{ab}\eta^{ac}\,\overline{\sigma}^{b} - \frac{\mathrm{i}}{2}\,\varepsilon^{abcd}\,\overline{\sigma}_{d}, \tag{C.39}$$

$$\sigma^{a}\,\overline{\sigma}^{bc} = -\frac{1}{2}\sum_{bc}\eta^{ab}\sigma^{c} + \frac{\mathrm{i}}{2}\,\varepsilon^{abcd}\,\sigma_{d},$$

$$\overline{\sigma}^{a}\sigma^{bc} = -\frac{1}{2}\sum_{bc}\eta^{ab}\,\overline{\sigma}^{c} - \frac{\mathrm{i}}{2}\,\varepsilon^{abcd}\,\overline{\sigma}_{d}, \tag{C.40}$$

105

$$\sigma^{ac}_{\alpha\beta}\,\sigma^{\ b}_{c\ \gamma\delta} = -\frac{1}{8}\sum_{\alpha\beta}\sum_{\gamma\delta}\varepsilon_{\alpha\gamma}\left(\sigma^a\overline{\sigma}^b\right)_{\beta\delta},$$

$$\sigma^{ac}_{\alpha\beta}\,\overline{\sigma}^{\ b}_{c\ \dot\alpha\dot\beta} = \frac{1}{8}\sum_{\alpha\beta}\sum_{\dot\alpha\dot\beta}\sigma^a_{\alpha\dot\alpha}\,\sigma^b_{\beta\dot\beta},$$

$$\overline{\sigma}^{ac}_{\dot\alpha\dot\beta}\,\overline{\sigma}^{\ b}_{c\ \dot\gamma\dot\delta} = \frac{1}{8}\sum_{\dot\alpha\dot\beta}\sum_{\dot\gamma\dot\delta}\varepsilon_{\dot\alpha\dot\gamma}\left(\overline{\sigma}^a\sigma^b\right)_{\dot\beta\dot\delta}, \tag{C.41}$$

$$\sigma^{ab}_{\alpha\beta}\,\sigma_{ab\,\gamma\delta} = \sum_{\alpha\beta}\varepsilon_{\alpha\gamma}\,\varepsilon_{\beta\delta},$$

$$\sigma^{ab}_{\alpha\beta}\,\overline{\sigma}_{ab\,\dot\alpha\dot\beta} = 0,$$

$$\overline{\sigma}^{ab}_{\dot\alpha\dot\beta}\,\overline{\sigma}_{ab\,\dot\gamma\dot\delta} = \sum_{\dot\alpha\dot\beta}\varepsilon_{\dot\alpha\dot\gamma}\,\varepsilon_{\dot\beta\dot\delta}, \tag{C.42}$$

$$\sigma^{ac}\,\sigma^{\ b}_{c} = \frac{3}{4}\,\eta^{ab} - \sigma^{ab},$$

$$\overline{\sigma}^{ac}\,\overline{\sigma}^{\ b}_{c} = \frac{3}{4}\,\eta^{ab} - \overline{\sigma}^{ab}, \tag{C.43}$$

$$\mathrm{tr}\left(\sigma^{ab}\sigma^{cd}\right) = -\frac{1}{2}\sum_{ab}\eta^{ac}\,\eta^{bd} - \frac{i}{2}\,\varepsilon^{abcd},$$

$$\mathrm{tr}\left(\overline{\sigma}^{ab}\,\overline{\sigma}^{cd}\right) = -\frac{1}{2}\sum_{ab}\eta^{ac}\,\eta^{bd} + \frac{i}{2}\,\varepsilon^{abcd}, \tag{C.44}$$

$$\left\{\sigma^{ab},\sigma^{cd}\right\}_{\alpha}^{\ \beta} = \mathrm{tr}\left(\sigma^{ab}\sigma^{cd}\right)\delta_{\alpha}^{\ \beta},$$

$$\left\{\overline{\sigma}^{ab},\overline{\sigma}^{cd}\right\}^{\dot\alpha}_{\ \dot\beta} = \mathrm{tr}\left(\overline{\sigma}^{ab}\,\overline{\sigma}^{cd}\right)\delta^{\dot\alpha}_{\ \dot\beta}, \tag{C.45}$$

$$\left[\sigma^{ab},\sigma^{cd}\right] = \sum_{ab}\sum_{cd}\eta^{ac}\,\sigma^{bd},$$

$$\left[\overline{\sigma}^{ab},\overline{\sigma}^{cd}\right] = \sum_{ab}\sum_{cd}\eta^{ac}\,\overline{\sigma}^{bd}. \tag{C.46}$$

Finally, we give some formulas for antisymmetric tensors. In spinor notation, $F_{[ab]}$ can be decomposed as follows:

$$F_{\alpha\dot\alpha\,\beta\dot\beta} = \varepsilon_{\dot\alpha\dot\beta}\,F_{(\alpha\beta)} + \varepsilon_{\alpha\beta}\,F_{(\dot\alpha\dot\beta)}. \tag{C.47}$$

This is equivalent to

$$F_{ab} = \frac{1}{2}\left(\sigma_{ab}\right)_{\alpha}^{\ \beta}F_{\beta}^{\ \alpha} - \frac{1}{2}\left(\overline{\sigma}_{ab}\right)^{\dot\alpha}_{\ \dot\beta}F^{\dot\beta}_{\ \dot\alpha}. \tag{C.48}$$

The dual tensor is defined as

$$\widetilde{F}^{[ab]} = \frac{1}{2}\,\varepsilon^{abcd}\,F_{cd}\,. \tag{C.49}$$

Its spinor decomposition reads

$$\widetilde{F}_{ab} = -\frac{\mathrm{i}}{2}\,(\sigma_{ab})_{\alpha}{}^{\beta}\,F_{\beta}{}^{\alpha} - \frac{\mathrm{i}}{2}\,(\overline{\sigma}_{ab})^{\dot{\alpha}}{}_{\dot{\beta}}\,F^{\dot{\beta}}{}_{\dot{\alpha}}\,, \tag{C.50}$$

$$\widetilde{F}_{\alpha\dot{\alpha}\,\beta\dot{\beta}} = -\mathrm{i}\,\varepsilon_{\dot{\alpha}\dot{\beta}}\,F_{\alpha\beta} + \mathrm{i}\,\varepsilon_{\alpha\beta}\,F_{\dot{\alpha}\dot{\beta}}\,. \tag{C.51}$$

If F is real, one has $(F_{\alpha\beta})^{*} = F_{\dot{\alpha}\dot{\beta}}$. If F is complex, it can be written as the sum of a self-dual and an anti-self-dual tensor. Self-duality means

$$F_{ab} = \mathrm{i}\,\widetilde{F}_{ab} \quad \Leftrightarrow \quad F_{\dot{\alpha}\dot{\beta}} = 0 \tag{C.52}$$

and anti-self-duality is defined by

$$F_{ab} = -\mathrm{i}\,\widetilde{F}_{ab} \quad \Leftrightarrow \quad F_{\alpha\beta} = 0\,. \tag{C.53}$$

Appendix D

Non-Linear Ricci Identities

Below we give the solution of the Ricci identities up to dimension 2, i. e., the non-linear versions of (7.1–4) and (7.6–13).

$$\mathcal{D}_\beta^E\,\overline{N}_{\alpha\dot\alpha}^{DCBA} = i\,\overline{N}_{(\beta\alpha)\dot\alpha}^{[EDCBA]} - \frac{i}{5}\sum_{DCBA}\left(V_{\beta\alpha}^{ED}\,\overline{W}_{\dot\alpha}^{CBA} - V_{\beta\alpha}^{DC}\,\overline{W}_{\dot\alpha}^{BAE}\right)$$

$$+ \frac{i}{4!}\,\varepsilon_{\beta\alpha}\sum_{DCBA}\left(8\,S^{ED}\,\overline{W}_{\dot\alpha}^{CBA} + 4\,\overline{W}_{\dot\alpha\dot\beta}^{ED}\,\overline{W}^{\dot\beta CBA}\right.$$

$$\left. - 6\,\overline{W}_{\dot\alpha\dot\beta}^{DC}\,\overline{W}^{\dot\beta BAE} + 2\,\overline{W}_{\dot\alpha}^{EDF}\,\overline{M}_F^{CBA} - 3\,\overline{W}_{\dot\alpha}^{DCF}\,\overline{M}_F^{BAE}\right) \qquad (D.1)$$

$$\mathcal{D}_{\dot\beta E}\,\overline{N}_{\alpha\dot\alpha}^{DCBA} = i\,\varepsilon_{\dot\beta\dot\alpha}\,\overline{M}_{\alpha E}^{[DCBA]} + \frac{i}{4!}\sum_{\dot\beta\dot\alpha}\sum_{DCBA}\left(4i\,\delta_E^D\,\mathcal{D}_{\alpha\dot\beta}\,\overline{W}_{\dot\alpha}^{CBA}\right.$$

$$\left. - 4\,U_{\alpha\dot\beta E}^D\,\overline{W}_{\dot\alpha}^{CBA} - \frac{3}{2}\,W_{\alpha EFG}\,\overline{W}_{\dot\beta}^{FDC}\,\overline{W}_{\dot\alpha}^{GBA}\right) \qquad (D.2)$$

$$\mathcal{D}_\alpha^E\,\overline{M}_D^{CBA} = \overline{M}_{\alpha D}^{ECBA} + i\,\delta_D^E\,\mathcal{D}_{\alpha\dot\alpha}\,\overline{W}^{\dot\alpha CBA} - 3\,U_{\alpha\dot\alpha D}^E\,\overline{W}^{\dot\alpha CBA}$$

$$+ \frac{1}{2}\sum_{CBA}\left(U_{\alpha\dot\alpha D}^C\,\overline{W}^{\dot\alpha EBA} + \delta_D^C\,U_{\alpha\dot\alpha F}^E\,\overline{W}^{\dot\alpha FBA}\right.$$

$$\left. - \frac{1}{2}\,W_{\alpha DFG}\,\overline{W}_{\dot\alpha}^{FEC}\,\overline{W}^{\dot\alpha GBA}\right) \qquad (D.3)$$

$$\mathcal{D}_{\dot\alpha E}\,\overline{M}_D^{CBA} = \frac{1}{3!}\sum_{CBA}\left[6\,\delta_D^C\,\overline{\Lambda}_{\dot\alpha E}^{BA} - 12\,\delta_E^C\,\overline{\Lambda}_{\dot\alpha D}^{BA}\right.$$

$$- 2\,\overline{V}_{\dot\alpha\dot\beta ED}\,\overline{W}^{\dot\beta CBA} + 2\left(\delta_E^C\,\overline{V}_{\dot\alpha\dot\beta DF} + \delta_D^C\,\overline{V}_{\dot\alpha\dot\beta EF}\right)\overline{W}^{\dot\beta FBA}$$

$$+ 2\,\overline{S}_{ED}\,\overline{W}_{\dot\alpha}^{CBA} + 6\left(\delta_E^C\,\overline{S}_{DF} - \delta_D^C\,\overline{S}_{EF}\right)\overline{W}_{\dot\alpha}^{FBA}$$

$$\left. - 3\,M_{EDF}^C\,\overline{W}_{\dot\alpha}^{FBA} - i\,W_{EDF}^\alpha\,\overline{N}_{\alpha\dot\alpha}^{FCBA}\right] \qquad (D.4)$$

$$\sum_{\dot\beta\dot\alpha}\sum_{DCBA}\left(\mathcal{D}_{\dot\beta}^\alpha\,\overline{N}_{\alpha\dot\alpha}^{DCBA} + 4\,\overline{W}_{\dot\beta\dot\alpha}^{DE}\,\overline{M}_E^{CBA} - 8\,\overline{W}_{\dot\beta\dot\alpha\dot\gamma}^D\,\overline{W}^{\dot\gamma CBA} - 16\,\overline{\Psi}_{\dot\beta}^D\,\overline{W}_{\dot\alpha}^{CBA}\right.$$

$$\left. + 12\,\overline{\Lambda}_{\dot\beta E}^{DC}\,\overline{W}_{\dot\alpha}^{EBA} + 2\,\overline{V}_{\dot\beta\dot\gamma EF}\,\overline{W}^{\dot\gamma EDC}\,\overline{W}_{\dot\alpha}^{FBA}\right) = 0 \qquad (D.5)$$

$$\mathcal{D}_\alpha^D \overline{\Lambda}_{\dot\alpha C}^{BA} = \frac{1}{12} \sum_{BA} \Big[3\mathrm{i}\, \mathcal{D}_{\alpha\dot\alpha} \overline{M}_C^{DBA} - 6\mathrm{i}\, \delta_C^D \, \mathcal{D}_\alpha^{\dot\beta} \, \overline{W}_{\dot\beta\dot\alpha}^{BA} + 4\mathrm{i}\, \delta_C^B \, \mathcal{D}_\alpha^{\dot\beta} \, \overline{W}_{\dot\beta\dot\alpha}^{DA}$$

$$+ 3\, U_{\alpha\dot\alpha E}^D \, \overline{M}_C^{EBA} + 6\, U_{\alpha C}^{D\dot\beta} \, \overline{W}_{\dot\beta\dot\alpha}^{BA} - 4\, \delta_C^B \, U_{\alpha E}^{D\dot\beta} \, \overline{W}_{\dot\beta\dot\alpha}^{EA}$$

$$+ \mathrm{i}\, \overline{V}_{\dot\alpha}^{\ \dot\beta}{}_{CE} \, \overline{N}_{\alpha\dot\beta}^{EDBA} + 3\mathrm{i}\, W_{\alpha}^{\ \beta}{}_{CE} \, \overline{N}_{\beta\dot\alpha}^{EDBA} - 3\mathrm{i}\, \overline{S}_{CE} \, \overline{N}_{\alpha\dot\alpha}^{EDBA}$$

$$+ 6\, \Psi_{\dot\alpha\dot\beta\alpha C} \, \overline{W}^{\dot\beta DBA} + 4\, \delta_C^B \, \Psi_{\dot\alpha\dot\beta\alpha E} \, \overline{W}^{\dot\beta EDA}$$

$$- 6\, \Psi_{\alpha C} \, \overline{W}_{\dot\alpha}^{DBA} + 12\, \delta_C^D \, \Psi_{\alpha E} \, \overline{W}_{\dot\alpha}^{EBA} - 12\, \delta_C^B \, \Psi_{\alpha E} \, \overline{W}_{\dot\alpha}^{EDA}$$

$$+ 3\, \Lambda_{\alpha CE}^D \, \overline{W}_{\dot\alpha}^{EBA} - 12\, \Lambda_{\alpha CE}^B \, \overline{W}_{\dot\alpha}^{EDA}$$

$$+ \big(\mathcal{D}_\alpha^D \, \overline{V}_{\dot\alpha\dot\beta CE} \big) \overline{W}^{\dot\beta EBA} + 3 \big(\mathcal{D}_\alpha^D \, \overline{S}_{CE} \big) \overline{W}_{\dot\alpha}^{EBA}$$

$$- 2\, V_{\alpha\beta}^{DE} \, W_{ECF}^\beta \, \overline{W}_{\dot\alpha}^{FBA} + 2\, V_{\alpha\beta}^{BE} \, W_{ECF}^\beta \, \overline{W}_{\dot\alpha}^{FDA}$$

$$+ 6\, \overline{W}_{\dot\alpha\dot\beta}^{BE} \, W_{\alpha ECF} \, \overline{W}^{\dot\beta FDA} + 6\, S^{BE} \, W_{\alpha ECF} \, \overline{W}_{\dot\alpha}^{FDA} \Big] \tag{D.6}$$

$$\mathcal{D}_{\dot\beta D} \overline{\Lambda}_{\dot\alpha C}^{BA} = \varepsilon_{\dot\beta\dot\alpha} P_{[DC]}^{[BA]} - \frac{1}{6}\, \varepsilon_{\dot\beta\dot\alpha} \widehat{\sum_{DC}} \sum_{BA} \delta_D^B \, \overline{V}_{CE}^{\dot\gamma\dot\delta} \, \overline{W}_{\dot\gamma\dot\delta}^{EA}$$

$$+ \frac{1}{24}\, \varepsilon_{\dot\beta\dot\alpha} \sum_{DC} \sum_{BA} \Big[\frac{\mathrm{i}}{4}\, \mathcal{D}_{\gamma\dot\gamma} \big(W_{DCE}^\gamma \overline{W}^{\dot\gamma BAE} \big) + \frac{1}{2}\, U_{\gamma\dot\gamma D}^E \, W_{ECF}^\gamma \, \overline{W}^{\dot\gamma BAF}$$

$$- \frac{1}{2}\, U_{\gamma\dot\gamma E}^B \, W_{DCF}^\gamma \, \overline{W}^{\dot\gamma EAF} - V_{\gamma\delta}^{BA} \, W_{DC}^{\gamma\delta} + \overline{V}_{DC}^{\dot\gamma\dot\delta} \, \overline{W}_{\dot\gamma\dot\delta}^{BA}$$

$$+ 3\, S^{BE} \, M_{EDC}^A - 3\, \overline{S}_{DE} \, \overline{M}_C^{EBA} + W_{DCE}^\gamma \, \mathcal{D}_\gamma^B S^{AE} - \overline{W}_{\dot\gamma}^{BAE} \, \mathcal{D}_D^{\dot\gamma} \overline{S}_{CE} \Big]$$

$$+ \frac{1}{72} \sum_{\dot\beta\dot\alpha} \sum_{BA} \Big[36\, \delta_D^B \, \overline{P}_{\dot\beta\dot\alpha C}^A - 12\, \delta_C^B \, \overline{P}_{\dot\beta\dot\alpha D}^A$$

$$- 24\, \overline{V}_{\dot\beta\dot\gamma DC} \, \overline{W}_{\dot\alpha}^{\dot\gamma BA} - 12 \big(\delta_D^B \, \overline{V}_{\dot\beta\dot\gamma CE} - \delta_C^B \, \overline{V}_{\dot\beta\dot\gamma DE} \big) \overline{W}_{\dot\alpha}^{\dot\gamma EA}$$

$$+ 36\, \overline{S}_{DC} \, \overline{W}_{\dot\beta\dot\alpha}^{BA} - 12 \big(3\, \delta_D^B \, \overline{S}_{CE} + \delta_C^B \, \overline{S}_{DE} \big) \overline{W}_{\dot\beta\dot\alpha}^{EA}$$

$$+ 9\, \overline{V}_{\dot\beta\dot\alpha DE} \, \overline{M}_C^{EBA} - 3\, \overline{V}_{\dot\beta\dot\alpha CE} \, \overline{M}_D^{EBA} - 18\, \overline{W}_{\dot\beta\dot\alpha}^{BE} \, M_{EDC}^A$$

$$+ 3\, \overline{V}_{\dot\beta\dot\alpha\dot\gamma DCE} \, \overline{W}^{\dot\gamma EBA} + 2 \big(5\, \mathcal{D}_{\dot\beta D} \overline{S}_{CE} + \mathcal{D}_{\dot\beta C} \overline{S}_{DE} \big) \overline{W}_{\dot\alpha}^{EBA}$$

$$+ 9\mathrm{i}\, W_{DCE}^\alpha \, \mathcal{D}_{\alpha\dot\beta} \overline{W}_{\dot\alpha}^{BAE} + 4\mathrm{i} \big(\mathcal{D}_{\alpha\dot\beta} W_{DCE}^\alpha \big) \overline{W}_{\dot\alpha}^{BAE}$$

$$- \big(5\, U_{\alpha\dot\beta D}^E \, W_{ECF}^\alpha + U_{\alpha\dot\beta C}^E \, W_{EDF}^\alpha \big) \overline{W}_{\dot\alpha}^{BAF} + U_{\alpha\dot\beta F}^E \, W_{EDC}^\alpha \, \overline{W}_{\dot\alpha}^{FBA} \Big],$$

$$P_{DC}^{BA} = \overline{P}_{DC}^{BA} \tag{D.7}$$

$$\frac{1}{2}\big[\mathcal{D}_{\dot\beta D},\mathcal{D}_{\dot\alpha C}\big]S^{BA} = S^{(BA)}_{(\dot\beta\dot\alpha)[DC]} + \varepsilon_{\dot\beta\dot\alpha}\,S^{(BA)}_{(DC)} - \frac{1}{4}\,\varepsilon_{\dot\beta\dot\alpha}\,\widehat{\sum_{DC}}\,\widehat{\sum_{BA}}\,\delta^B_D\big(\mathrm{i}\,\mathcal{D}^{\gamma\dot\gamma}U^A_{\gamma\dot\gamma C}$$

$$- V^{AE}_{\gamma\delta}\,W^{\gamma\delta}_{EC} + \overline{V}^{\dot\gamma\dot\delta}_{CE}\,\overline{W}^{EA}_{\dot\gamma\dot\delta}\big),$$

$$S^{BA}_{DC} = \overline{S}^{BA}_{DC} \tag{D.8}$$

$$\frac{1}{2}\big[\mathcal{D}_{\dot\beta D},\mathcal{D}_{\dot\alpha C}\big]V^{BA}_{\beta\alpha} = V^{[BA]}_{(\beta\alpha)(\dot\beta\dot\alpha)[DC]} + \frac{1}{16}\sum_{\beta\alpha}\sum_{\dot\beta\dot\alpha}\sum_{DC}\sum_{BA}\Big[4\mathrm{i}\,\delta^B_D\,\mathcal{D}_{\beta\dot\beta}U^A_{\alpha\dot\alpha C}$$

$$- \frac{\mathrm{i}}{2}\,\mathcal{D}_{\beta\dot\beta}\big(W_{\alpha DCE}\,\overline{W}^{BAE}_{\dot\alpha}\big) + U^E_{\beta\dot\beta D}\,W_{\alpha ECF}\,\overline{W}^{BAF}_{\dot\alpha}$$

$$- U^B_{\beta\dot\beta E}\,W_{\alpha DCF}\,\overline{W}^{EAF}_{\dot\alpha}\Big] - \varepsilon_{\dot\beta\dot\alpha}\,\overline{S}^{BA}_{\beta\alpha DC}$$

$$+ \frac{1}{2}\,\varepsilon_{\dot\beta\dot\alpha}\sum_{\beta\alpha}\widehat{\sum_{DC}}\sum_{BA}\Big(\delta^B_D\,\rho^A_{\beta\alpha C} + \mathrm{i}\,\delta^B_D\,\mathcal{D}_\beta{}^{\dot\gamma}U^A_{\alpha\dot\gamma C}$$

$$- U^B_{\beta D}{}^{\dot\gamma}U^A_{\alpha\dot\gamma C} - \delta^B_D\,V^{AE}_{\beta\gamma}\,W^\gamma{}_{\alpha EC} + \delta^B_D\,S^{AE}\,W_{\beta\alpha EC}$$

$$- \frac{1}{4}\,W_{\beta\alpha DE}\,\overline{M}^{EBA}_C + \frac{1}{4}\,U^E_{\beta D}{}^{\dot\gamma}W_{\alpha ECF}\,\overline{W}^{BAF}_{\dot\gamma}\Big),$$

$$V^{BA}_{\beta\alpha\dot\beta\dot\alpha DC} = \overline{V}^{BA}_{\beta\alpha\dot\beta\dot\alpha DC} \tag{D.9}$$

$$\mathcal{D}^D_\delta V^{CBA}_{\gamma\beta\alpha} = V^{[DCBA]}_{(\delta\gamma\beta\alpha)} + \frac{1}{24}\sum_{\gamma\beta\alpha}\sum_{CBA}\varepsilon_{\delta\gamma}\Big(-\mathcal{D}^C_\beta\mathcal{D}^B_\alpha S^{AD} + \frac{1}{6}\,\mathcal{D}_\beta{}^{\dot\alpha}\,\overline{N}^{DCBA}_{\alpha\dot\alpha}$$

$$+ \frac{\mathrm{i}}{3}\,U^{D\dot\alpha}_{\beta E}\,\overline{N}^{ECBA}_{\alpha\dot\alpha} + \mathrm{i}\,U^{C\dot\alpha}_{\beta E}\,\overline{N}^{EBAD}_{\alpha\dot\alpha} - 6\,V^{DC}_{\beta e}\,V^{e\,BA}_\alpha$$

$$- 6\,S^{DC}V^{BA}_{\beta\alpha} - \frac{1}{6}\,V^{DE}_{\beta\alpha}\,\overline{M}^{CBA}_E - \frac{3}{2}\,V^{CE}_{\beta\alpha}\,\overline{M}^{BAD}_E$$

$$- \overline{\Psi}^D_{\beta\alpha\dot\alpha}\,\overline{W}^{\dot\alpha CBA} - \overline{\Psi}^C_{\beta\alpha\dot\alpha}\,\overline{W}^{\dot\alpha BAD} - \frac{3}{2}\,\overline{W}^{DCE}_{\dot\alpha}\,\mathcal{D}^{\dot\alpha}_E V^{BA}_{\beta\alpha}$$

$$- \frac{1}{2}\,\overline{W}^{CBE}_{\dot\alpha}\,\mathcal{D}^{\dot\alpha}_E V^{AD}_{\beta\alpha} - W_{\beta\alpha EF}\,\overline{W}^{EDC}_{\dot\alpha}\,\overline{W}^{\dot\alpha FBA}\Big) \tag{D.10}$$

$$\mathcal{D}^F_\gamma\,\overline{N}^{EDCBA}_{\beta\alpha\dot\alpha} = \overline{N}^{[FEDCBA]}_{(\gamma\beta\alpha)\dot\alpha} - \frac{4}{3!\,5!}\sum_{\gamma\beta\alpha}\sum_{EDCBA}\Big(V^{FED}_{\gamma\beta\alpha}\,\overline{W}^{CBA}_{\dot\alpha} + V^{EDC}_{\gamma\beta\alpha}\,\overline{W}^{BAF}_{\dot\alpha}$$

$$+ \frac{8}{3}\mathrm{i}\,V^{FE}_{\gamma\beta}\,\overline{N}^{DCBA}_{\alpha\dot\alpha} + \frac{8}{3}\mathrm{i}\,V^{ED}_{\gamma\beta}\,\overline{N}^{CBAF}_{\alpha\dot\alpha}\Big)$$

$$+ \frac{1}{3}\frac{1}{5!}\sum_{\beta\alpha}\sum_{EDCBA}\varepsilon_{\gamma\beta}\Big[-10\mathrm{i}\,V^{\delta FE}_\alpha\,\overline{N}^{DCBA}_{\delta\dot\alpha} - 8\mathrm{i}\,V^{\delta ED}_\alpha\,\overline{N}^{CBAF}_{\delta\dot\alpha}$$

$$- 10\mathrm{i}\,\overline{W}^{\dot\beta FE}_{\dot\alpha}\,\overline{N}^{DCBA}_{\alpha\dot\beta} - 10\mathrm{i}\,\overline{W}^{\dot\beta ED}_{\dot\alpha}\,\overline{N}^{CBAF}_{\alpha\dot\beta} + 30\mathrm{i}\,S^{FE}\,\overline{N}^{DCBA}_{\alpha\dot\alpha}$$

$$- 10\mathrm{i}\,\overline{M}^{FED}_G\,\overline{N}^{GCBA}_{\alpha\dot\alpha} - 5\mathrm{i}\,\overline{M}^{EDC}_G\,\overline{N}^{GBAF}_{\alpha\dot\alpha} - 5\,\overline{W}^{FEG}_{\dot\alpha}\,\overline{M}^{DCBA}_{\alpha G}$$

$$- 5\,\overline{W}_{\dot{\alpha}}^{EDG}\,\overline{M}_{\alpha G}^{CBAF} + 16\,\overline{W}_{\dot{\alpha}}^{EDC}\,\mathcal{D}_{\alpha}^{B}\,S^{AF} + 5\mathrm{i}\,\overline{W}^{\dot{\beta}\,FED}\,\Big(\mathcal{D}_{\alpha\dot{\beta}}\,\overline{W}_{\dot{\alpha}}^{CBA}$$

$$+ 3\,\mathcal{D}_{\alpha\dot{\alpha}}\,\overline{W}_{\dot{\beta}}^{CBA}\Big) - \mathrm{i}\,\overline{W}^{\dot{\beta}\,EDC}\,\Big(\mathcal{D}_{\alpha\dot{\beta}}\,\overline{W}_{\dot{\alpha}}^{BAF} - 11\,\mathcal{D}_{\alpha\dot{\alpha}}\,\overline{W}_{\dot{\beta}}^{BAF}\Big)$$

$$- 4\,\Big(U_{\alpha G}^{F\dot{\beta}}\,\overline{W}_{\dot{\beta}}^{GED} + 4\,U_{\alpha G}^{E\dot{\beta}}\,\overline{W}_{\dot{\beta}}^{GFD}\Big)\,\overline{W}_{\dot{\alpha}}^{CBA} + 5\,\Big(U_{\alpha G}^{F\dot{\beta}}\,\overline{W}_{\dot{\alpha}}^{GED}$$

$$+ 4\,U_{\alpha G}^{E\dot{\beta}}\,\overline{W}_{\dot{\alpha}}^{GFD}\Big)\,\overline{W}_{\dot{\beta}}^{CBA} - 15\,U_{\alpha G}^{E\dot{\beta}}\,\overline{W}_{\dot{\alpha}}^{GDC}\,\overline{W}_{\dot{\beta}}^{BAF}$$

$$- \frac{15}{2}\,W_{\alpha GHK}\,\overline{W}_{\dot{\alpha}}^{GED}\,\overline{W}_{\dot{\beta}}^{HCB}\,\overline{W}^{\dot{\beta}\,KAF}\Big] \tag{D.11}$$

$$\mathcal{D}_{\dot{\beta}F}\,\overline{N}_{\beta\alpha\dot{\alpha}}^{EDCBA} = \varepsilon_{\dot{\beta}\dot{\alpha}}\,\overline{M}_{(\beta\alpha)F}^{[EDCBA]} - \frac{1}{48}\sum_{\beta\alpha}\sum_{\dot{\beta}\dot{\alpha}}\sum_{EDCBA}\Big(\delta_{F}^{E}\,\mathcal{D}_{\beta\dot{\beta}}\,\overline{N}_{\alpha\dot{\alpha}}^{DCBA}$$

$$+ 2\mathrm{i}\,U_{\beta\dot{\beta}F}^{E}\,\overline{N}_{\alpha\dot{\alpha}}^{DCBA} + \frac{24}{5}\,\delta_{F}^{E}\,V_{\beta\alpha}^{DC}\,\overline{W}_{\dot{\beta}\dot{\alpha}}^{BA} + 8\,\delta_{F}^{E}\,\overline{\Psi}_{\beta\alpha\dot{\beta}}^{D}\,\overline{W}_{\dot{\alpha}}^{CBA}$$

$$- \frac{4}{5}\,\overline{W}_{\dot{\beta}}^{EDC}\,\mathcal{D}_{\dot{\alpha}F}V_{\beta\alpha}^{BA} + 2\,W_{\beta\alpha FG}\,\overline{W}_{\dot{\beta}}^{GED}\,\overline{W}_{\dot{\alpha}}^{CBA}$$

$$+ 3\,\delta_{F}^{E}\,W_{\beta\alpha GH}\,\overline{W}_{\dot{\beta}}^{GDC}\,\overline{W}_{\dot{\alpha}}^{HBA} + \mathrm{i}\,W_{\beta FGH}\,\overline{W}_{\dot{\beta}}^{GED}\,\overline{N}_{\alpha\dot{\alpha}}^{HCBA}\Big) \tag{D.12}$$

$$\mathcal{D}_{\beta}^{F}\,\overline{M}_{\alpha E}^{DCBA} = -\overline{M}_{\beta\alpha E}^{FDCBA} - \delta_{E}^{F}\,\mathcal{D}_{\beta}^{\dot{\alpha}}\,\overline{N}_{\alpha\dot{\alpha}}^{DCBA} - 3\mathrm{i}\,U_{\beta E}^{F\dot{\alpha}}\,\overline{N}_{\alpha\dot{\alpha}}^{DCBA} - \mathrm{i}\,U_{\alpha E}^{F\dot{\alpha}}\,\overline{N}_{\beta\dot{\alpha}}^{DCBA}$$

$$+ \frac{1}{4!}\sum_{DCBA}\Big[-4\mathrm{i}\,U_{\beta E}^{D\dot{\alpha}}\,\overline{N}_{\alpha\dot{\alpha}}^{CBAF} + 4\mathrm{i}\,\delta_{E}^{D}\,U_{\beta G}^{F\dot{\alpha}}\,\overline{N}_{\alpha\dot{\alpha}}^{GCBA}$$

$$+ \frac{24}{5}\,\Big(V_{\beta\alpha}^{FD}\,\overline{M}_{E}^{CBA} - V_{\beta\alpha}^{DC}\,\overline{M}_{E}^{BAF}\Big)$$

$$+ \frac{12}{5}\,\Big(\overline{W}_{\dot{\alpha}}^{FDC}\,\mathcal{D}_{E}^{\dot{\alpha}}V_{\beta\alpha}^{BA} - \overline{W}_{\dot{\alpha}}^{DCB}\,\mathcal{D}_{E}^{\dot{\alpha}}V_{\beta\alpha}^{AF}\Big)$$

$$- 3\mathrm{i}\,W_{\beta EGH}\,\overline{W}^{\dot{\alpha}\,GFD}\,\overline{N}_{\alpha\dot{\alpha}}^{HCBA} + \mathrm{i}\,W_{\alpha EGH}\,\overline{W}^{\dot{\alpha}\,GFD}\,\overline{N}_{\beta\dot{\alpha}}^{HCBA}$$

$$+ \varepsilon_{\beta\alpha}\Big(6\,\delta_{E}^{F}\,\overline{W}_{\dot{\alpha}\dot{\beta}}^{DC}\,\overline{W}^{\dot{\alpha}\dot{\beta}\,BA} - 8\,S^{FD}\,\overline{M}_{E}^{CBA} - 2\,\overline{M}_{E}^{FDG}\,\overline{M}_{G}^{CBA}$$

$$+ 3\,\overline{M}_{E}^{DCG}\,\overline{M}_{G}^{BAF} - 12\,\overline{\Lambda}_{\dot{\alpha}E}^{FD}\,\overline{W}^{\dot{\alpha}\,CBA} + 12\,\overline{\Lambda}_{\dot{\alpha}E}^{DC}\,\overline{W}^{\dot{\alpha}\,BAF}$$

$$+ 6\,\delta_{E}^{F}\,\overline{\Lambda}_{\dot{\alpha}G}^{DC}\,\overline{W}^{\dot{\alpha}\,GBA} + 12\,\delta_{E}^{D}\,\overline{\Lambda}_{\dot{\alpha}G}^{CF}\,\overline{W}^{\dot{\alpha}\,GBA} - 12\,\delta_{E}^{D}\,\overline{\Lambda}_{\dot{\alpha}G}^{CB}\,\overline{W}^{\dot{\alpha}\,GAF}$$

$$- 4\,\overline{W}_{\dot{\alpha}}^{DCB}\,\mathcal{D}_{E}^{\dot{\alpha}}\,S^{AF} - 2\,\overline{V}_{EG}^{\dot{\alpha}\dot{\beta}}\,\overline{W}_{\dot{\alpha}}^{GDC}\,\overline{W}_{\dot{\beta}}^{BAF}$$

$$+ \delta_{E}^{F}\,\overline{V}_{GH}^{\dot{\alpha}\dot{\beta}}\,\overline{W}_{\dot{\alpha}}^{GDC}\,\overline{W}_{\dot{\beta}}^{HBA} + 8\,\overline{S}_{EG}\,\overline{W}_{\dot{\alpha}}^{GFD}\,\overline{W}^{\dot{\alpha}\,CBA}$$

$$- 6\,\overline{S}_{EG}\,\overline{W}_{\dot{\alpha}}^{GDC}\,\overline{W}^{\dot{\alpha}\,BAF} - 3\,\delta_{E}^{F}\,\overline{S}_{GH}\,\overline{W}_{\dot{\alpha}}^{GDC}\,\overline{W}^{\dot{\alpha}\,HBA}$$

$$+ 6\,M_{EGH}^{D}\,\overline{W}_{\dot{\alpha}}^{GCB}\,\overline{W}^{\dot{\alpha}\,HAF} + \frac{3}{2}\mathrm{i}\,W_{EGH}^{\gamma}\,\overline{W}^{\dot{\alpha}\,GDC}\,\overline{N}_{\gamma\dot{\alpha}}^{HBAF}\Big)\Big] \tag{D.13}$$

111

$$\mathcal{D}_{\dot\alpha F}\,\overline{M}_{\alpha E}^{DCBA} = \frac{1}{6}\sum_{DCBA}\Big[\, \mathrm{i}\,\delta_E^D\,\mathcal{D}_{\alpha\dot\alpha}\,\overline{M}_F^{CBA} - 2\mathrm{i}\,\delta_F^D\,\mathcal{D}_{\alpha\dot\alpha}\,\overline{M}_E^{CBA} + 6\mathrm{i}\,\delta_F^D\,\delta_E^C\,\mathcal{D}_\alpha^{\;\dot\beta}\,\overline{W}_{\dot\beta\dot\alpha}^{BA}$$

$$+\, 2\,U_{\alpha\dot\alpha\,F}^{D}\,\overline{M}_E^{CBA} - U_{\alpha\dot\alpha\,E}^{D}\,\overline{M}_F^{CBA} + 6\,\delta_F^D\,U_{\alpha E}^{C\dot\beta}\,\overline{W}_{\dot\beta\dot\alpha}^{BA}$$

$$-\,\frac{\mathrm{i}}{2}\,\overline{V}_{\dot\alpha\;FE}^{\;\dot\beta}\,\overline{N}_{\alpha\dot\beta}^{DCBA} + \mathrm{i}\,\delta_E^D\,\overline{V}_{\dot\alpha\;FG}^{\;\dot\beta}\,\overline{N}_{\alpha\dot\beta}^{GCBA}$$

$$-\,\frac{\mathrm{i}}{2}\,W_{\alpha\;FE}^{\;\beta}\,\overline{N}_{\beta\dot\alpha}^{DCBA} + \mathrm{i}\,\big(\delta_E^D\,W_{\alpha\;FG}^{\;\beta} - 2\,\delta_F^D\,W_{\alpha\;EG}^{\;\beta}\big)\,\overline{N}_{\beta\dot\alpha}^{GCBA}$$

$$-\,\frac{\mathrm{i}}{2}\,\overline{S}_{FE}\,\overline{N}_{\alpha\dot\alpha}^{DCBA} + \mathrm{i}\,\delta_E^D\,\overline{S}_{FG}\,\overline{N}_{\alpha\dot\alpha}^{GCBA} + \mathrm{i}\,M_{FEG}^{D}\,\overline{N}_{\alpha\dot\alpha}^{GCBA}$$

$$+\,2\,\big(2\,\delta_E^D\,\Psi_{\dot\alpha\dot\beta\,\alpha F} - 3\,\delta_F^D\,\Psi_{\dot\alpha\dot\beta\,\alpha E}\big)\,\overline{W}^{\dot\beta CBA} + 6\,\delta_F^D\,\delta_E^C\,\Psi_{\dot\alpha\dot\beta\,\alpha G}\,\overline{W}^{\dot\beta GBA}$$

$$+\,6\,\delta_F^D\,\Psi_{\alpha E}\,\overline{W}_{\dot\alpha}^{CBA} - 18\,\delta_F^D\,\delta_E^C\,\Psi_{\alpha G}\,\overline{W}_{\dot\alpha}^{GBA}$$

$$+\,3\,\Lambda_{\alpha FE}^{D}\,\overline{W}_{\dot\alpha}^{CBA} + 6\,\big(\delta_E^D\,\Lambda_{\alpha FG}^{C} - 2\,\delta_F^D\,\Lambda_{\alpha EG}^{C}\big)\,\overline{W}_{\dot\alpha}^{GBA}$$

$$-\,\big(\mathcal{D}_\alpha^D\,\overline{V}_{\dot\alpha\dot\beta\,FE}\big)\,\overline{W}^{\dot\beta CBA} + \big(\mathcal{D}_\alpha^D\,\overline{S}_{FE}\big)\,\overline{W}_{\dot\alpha}^{CBA} - \frac{1}{4}\,W_{FEG}^{\beta}\,\overline{N}_{\beta\alpha\dot\alpha}^{GDCBA}$$

$$-\,\frac{6}{5}\,V_{\alpha\beta}^{DC}\,W_{FEG}^{\beta}\,\overline{W}_{\dot\alpha}^{GBA} - \frac{1}{5}\,V_{\alpha\beta}^{DG}\,W_{GFE}^{\beta}\,\overline{W}_{\dot\alpha}^{CBA}$$

$$+\,V_{\alpha\beta}^{CG}\,\big(2\,\delta_F^D\,W_{GEH}^{\beta} - \delta_E^D\,W_{GFH}^{\beta}\big)\,\overline{W}_{\dot\alpha}^{HBA}$$

$$-\,2\,\overline{W}_{\dot\alpha\dot\beta}^{DG}\,W_{\alpha GFE}\,\overline{W}^{\dot\beta CBA} + 3\,\overline{W}_{\dot\alpha\dot\beta}^{CG}\,\big(\delta_F^D\,W_{\alpha GEH}$$

$$-\,\delta_E^D\,W_{\alpha GFH}\big)\,\overline{W}^{\dot\beta HBA} - S^{DG}\,W_{\alpha GFE}\,\overline{W}_{\dot\alpha}^{CBA}$$

$$+\,3\,S^{CG}\,\big(2\,\delta_F^D\,W_{\alpha GEH} - \delta_E^D\,W_{\alpha GFH}\big)\,\overline{W}_{\dot\alpha}^{HBA}$$

$$+\,\frac{3}{4}\,\big(2\,\overline{M}_E^{DCG}\,W_{\alpha GFH} - \overline{M}_F^{DCG}\,W_{\alpha GEH}\big)\,\overline{W}_{\dot\alpha}^{HBA}$$

$$+\,\frac{1}{2}\,W_{\alpha FEG}\,\overline{W}_{\dot\alpha}^{GDH}\,\overline{M}_H^{CBA} - \frac{3}{4}\,W_{\alpha FEG}\,\overline{W}_{\dot\alpha}^{DCH}\,\overline{M}_H^{GBA}$$

$$+\,\frac{3}{4}\mathrm{i}\,N_{\alpha\dot\beta FEGH}\,\overline{W}^{\dot\beta GDC}\,\overline{W}_{\dot\alpha}^{HBA}\,\Big] \tag{D.14}$$

$$\sum_{FED} \sum_{CBA} \left[12\, \delta_F^C\, P_{ED}^{BA} - 2\, \delta_F^C \left(V_{\alpha\beta}^{BA}\, W_{ED}^{\alpha\beta} + \overline{V}_{ED}^{\dot\alpha\dot\beta}\, \overline{W}_{\dot\alpha\dot\beta}^{BA} \right) \right.$$

$$+ 6\, \delta_F^C \left(S^{BG}\, M_{GED}^A + \overline{S}_{EG}\, \overline{M}_D^{GBA} \right)$$

$$+ \frac{1}{3}\, M_{FED}^G\, \overline{M}_G^{CBA} - 3\, M_{FEG}^C\, \overline{M}_D^{BAG} - \frac{1}{3}\, N_{FEDG}^{\alpha\dot\alpha}\, \overline{N}_{\alpha\dot\alpha}^{CBAG}$$

$$- W_{FEG}^\alpha\, \overline{M}_{\alpha D}^{GCBA} - \overline{W}_{\dot\alpha}^{CBG}\, M_{GFED}^{\dot\alpha A}$$

$$+ 2\, \delta_F^C \left(W_{EDG}^\alpha\, \mathcal{D}_\alpha^B\, S^{AG} + \overline{W}_{\dot\alpha}^{BAG}\, \mathcal{D}_E^{\dot\alpha}\, \overline{S}_{DG} \right)$$

$$- \frac{i}{3} \left(W_{FED}^\alpha\, \mathcal{D}_{\alpha\dot\alpha}\, \overline{W}^{\dot\alpha CBA} + \overline{W}^{\dot\alpha CBA}\, \mathcal{D}_{\alpha\dot\alpha}\, W_{FED}^\alpha \right)$$

$$+ \frac{i}{2}\, \delta_F^C \left(W_{EDG}^\alpha\, \mathcal{D}_{\alpha\dot\alpha}\, \overline{W}^{\dot\alpha BAG} + \overline{W}^{\dot\alpha BAG}\, \mathcal{D}_{\alpha\dot\alpha}\, W_{EDG}^\alpha \right)$$

$$- 6\, U_{\alpha\dot\alpha\,F}^C\, W_{EDG}^\alpha\, \overline{W}^{\dot\alpha BAG} + U_{\alpha\dot\alpha\,G}^C\, W_{FED}^\alpha\, \overline{W}^{\dot\alpha GBA}$$

$$+ U_{\alpha\dot\alpha\,F}^G\, W_{GED}^\alpha\, \overline{W}^{\dot\alpha CBA} - \delta_F^C \left(U_{\alpha\dot\alpha\,G}^B\, W_{EDH}^\alpha\, \overline{W}^{\dot\alpha GAH} \right.$$

$$\left. \left. + U_{\alpha\dot\alpha\,E}^G\, W_{GDH}^\alpha\, \overline{W}^{\dot\alpha BAH} + U_{\alpha\dot\alpha\,H}^G\, W_{GED}^\alpha\, \overline{W}^{\dot\alpha HBA} \right) \right] = 0 \tag{D.15}$$

Appendix E

Second Bianchi Identity

This appendix contains the solution of the second Bianchi identity (2.27) at dimensions $\frac{5}{2}$ and 3. (The solution at dimensions $\frac{3}{2}$ and 2 is part of the preceding appendix.) The following equations are the non-linear extensions of (7.17–23).

$$
\mathcal{D}^A_\gamma P_{\beta\alpha\dot\beta\dot\alpha} = \frac{1}{4} \sum_{\beta\alpha} \sum_{\dot\beta\dot\alpha} \Big[2\mathrm{i}\, \mathcal{D}_{\gamma\dot\beta}\, \overline{\Psi}^A_{\beta\alpha\dot\alpha} - \mathrm{i}\, \mathcal{D}_{\beta\dot\beta}\, \overline{\Psi}^A_{\gamma\alpha\dot\alpha} + 3\, U^A_{\gamma\dot\beta B}\, \overline{\Psi}^B_{\beta\alpha\dot\alpha}
$$
$$
+ V^{AB}_{\gamma\beta}\, \Psi_{\dot\beta\dot\alpha\alpha B} + 2\, V^{AB}_{\beta\alpha}\, \Psi_{\dot\beta\dot\alpha\gamma B} + 2\, \overline{W}^{AB}_{\dot\beta\dot\alpha}\, W_{\gamma\beta\alpha B}
$$
$$
+ \varepsilon_{\gamma\beta} \big(3\mathrm{i}\, \mathcal{D}_{\alpha\dot\beta}\, \overline{\Psi}^A_{\dot\alpha} - 2\, U^{A\dot\gamma}_{\alpha B}\, \overline{W}^B_{\dot\gamma\dot\beta\dot\alpha} - 4\, U^A_{\alpha\dot\beta B}\, \overline{\Psi}^B_{\dot\alpha}
$$
$$
- \overline{W}^{AB}_{\dot\beta\dot\gamma}\, \Psi^{\dot\gamma}_{\dot\alpha\alpha B} - \overline{W}^{AB}_{\dot\beta\dot\alpha}\, \Psi_{\alpha B} - 5\, S^{AB}\, \Psi_{\dot\beta\dot\alpha\alpha B} \big) \Big] \tag{E.1}
$$

$$
\mathcal{D}^A_\varepsilon W_{\delta\gamma\beta\alpha} = \frac{1}{12} \sum_{\delta\gamma\beta\alpha} \Big[2\, V^{AB}_{\varepsilon\delta}\, W_{\gamma\beta\alpha B} + V^{AB}_{\delta\gamma}\, W_{\varepsilon\beta\alpha B}
$$
$$
+ \varepsilon_{\varepsilon\delta} \big(\mathrm{i}\, \mathcal{D}^{\dot\alpha}_\gamma\, \overline{\Psi}^A_{\beta\alpha\dot\alpha} - U^{A\dot\alpha}_{\gamma B}\, \overline{\Psi}^B_{\beta\alpha\dot\alpha} - 2\, V^{AB}_{\gamma\beta}\, \Psi_{\alpha B} - 4\, S^{AB}\, W_{\gamma\beta\alpha B} \big) \Big] \tag{E.2}
$$

$$
\mathcal{D}^A_\alpha \overline{W}_{\dot\delta\dot\gamma\dot\beta\dot\alpha} = \frac{1}{12} \sum_{\dot\delta\dot\gamma\dot\beta\dot\alpha} \big(\mathrm{i}\, \mathcal{D}_{\alpha\dot\delta}\, \overline{W}^A_{\dot\gamma\dot\beta\dot\alpha} + 3\, U^A_{\alpha\dot\delta B}\, \overline{W}^B_{\dot\gamma\dot\beta\dot\alpha} + 2\, \overline{W}^{AB}_{\dot\delta\dot\gamma}\, \Psi_{\dot\beta\dot\alpha\alpha B} \big) \tag{E.3}
$$

$$
\mathcal{D}^A_\alpha R = -12\mathrm{i}\, \mathcal{D}^{\dot\alpha}_\alpha\, \overline{\Psi}^A_{\dot\alpha} + 6\, U^{\beta\dot\alpha A}_{B}\, \overline{\Psi}^B_{\beta\alpha\dot\alpha} - 18\, U^{A\dot\alpha}_{\alpha B}\, \overline{\Psi}^B_{\dot\alpha}
$$
$$
+ 6\, V^{\beta\gamma AB}\, W_{\beta\gamma\alpha B} + 12\, V^{AB}_{\alpha\beta}\, \Psi^\beta_B + 2\, \overline{W}^{AB}_{\dot\alpha\dot\beta}\, \Psi^{\dot\alpha\dot\beta}_{\alpha B} - 36\, S^{AB}\, \Psi_{\alpha B} \tag{E.4}
$$

$$
\mathcal{D}^C_\gamma \rho^B_{\beta\alpha A} = \frac{1}{3!} \sum_{\gamma\beta\alpha} \Big(2\mathrm{i}\, \delta^C_A\, \mathcal{D}^{\dot\alpha}_\gamma\, \overline{\Psi}^B_{\beta\alpha\dot\alpha} - \frac{\mathrm{i}}{2}\, \delta^B_A\, \mathcal{D}^{\dot\alpha}_\gamma\, \overline{\Psi}^C_{\beta\alpha\dot\alpha}
$$
$$
- 2\, \delta^C_A\, U^{B\dot\alpha}_{\gamma D}\, \overline{\Psi}^D_{\beta\alpha\dot\alpha} + \frac{1}{2}\, \delta^B_A\, U^{C\dot\alpha}_{\gamma D}\, \overline{\Psi}^D_{\beta\alpha\dot\alpha}
$$
$$
- 2\, \delta^C_A\, V^{\delta\, BD}_\gamma\, W_{\delta\beta\alpha D} - 4\, \delta^C_A\, V^{BD}_{\gamma\beta}\, \Psi_{\alpha D} + \delta^B_A\, V^{CD}_{\gamma\beta}\, \Psi_{\alpha D}
$$
$$
+ 2\, S^{CB}\, W_{\gamma\beta\alpha A} - 2\, \delta^C_A\, S^{BD}\, W_{\gamma\beta\alpha D} - \overline{M}^{CBD}_A\, W_{\gamma\beta\alpha D} \Big)
$$
$$
+ \frac{1}{2} \sum_{\beta\alpha} \Big[\mathrm{i}\, \mathcal{D}^{\dot\alpha}_\beta \big(W_{\gamma\alpha AD}\, \overline{W}^{DCB}_{\dot\alpha} \big) + 2\, V^{\delta\, CB}_\gamma\, W_{\delta\beta\alpha A}
$$
$$
- 4\, V^{CD}_{\gamma\beta}\, \Lambda^B_{\alpha DA} + U^{C\dot\alpha}_{\gamma D}\, W_{\beta\alpha AE}\, \overline{W}^{EDB}_{\dot\alpha} + \frac{2}{3}\, V^{CD}_{\gamma\beta}\, V^{BE}_{\alpha\delta}\, W^\delta_{EDA}
$$

$$+ 2 V_{\gamma\beta}^{CD} S^{BE} W_{\alpha EDA} - W_{\gamma ADE} \overline{W}^{\dot\alpha DCB} \overline{\Psi}_{\beta\alpha\dot\alpha}^{E}$$

$$+ i\, \varepsilon_{\gamma\beta} \mathcal{D}_{\alpha}{}^{\dot\alpha} \left(2\,\overline{\Lambda}_{\dot\alpha A}^{CB} - \frac{1}{3} \overline{V}_{\dot\alpha\dot\beta AD} \overline{W}^{\dot\beta DCB} - \overline{S}_{AD} \overline{W}_{\dot\alpha}^{DCB} \right)$$

$$+ \varepsilon_{\gamma\beta} \left(-\frac{1}{3} \delta_A^B V^{\delta\varepsilon CD} W_{\delta\varepsilon\alpha D} - 2\,\overline{W}_{\dot\alpha\dot\beta}^{CB} \Psi_{\alpha A}^{\dot\alpha\dot\beta} \right.$$

$$- \frac{2}{3} \delta_A^C \overline{W}_{\dot\alpha\dot\beta}^{BD} \Psi_{\alpha D}^{\dot\alpha\dot\beta} + \frac{2}{3} \delta_A^B \overline{W}_{\dot\alpha\dot\beta}^{CD} \Psi_{\alpha D}^{\dot\alpha\dot\beta} + 4\, S^{CD} \Lambda_{\alpha DA}^{B}$$

$$+ 2\,\overline{M}_A^{CBD} \Psi_{\alpha D} - \overline{W}_{\dot\alpha\dot\beta}^{CD} \overline{W}^{\dot\alpha\dot\beta BE} W_{\alpha EDA}$$

$$\left. - \frac{2}{3} S^{CD} V_{\alpha\delta}^{BE} W_{EDA}^{\delta} - 2\, S^{CD} S^{BE} W_{\alpha EDA} \right) \bigg] \tag{E.5}$$

$$\mathcal{D}_\alpha^C \overline{\rho}_{\dot\beta\dot\alpha A}^{B} = \frac{1}{2} \sum_{\dot\beta\dot\alpha} \bigg[-2i\, \delta_A^C \mathcal{D}_\alpha{}^{\dot\gamma} \overline{W}_{\dot\gamma\dot\beta\dot\alpha}^{B} + \frac{i}{2} \delta_A^B \mathcal{D}_\alpha{}^{\dot\gamma} \overline{W}_{\dot\gamma\dot\beta\dot\alpha}^{C}$$

$$+ i\, \mathcal{D}_{\alpha\dot\beta} \left(-2\,\overline{\Lambda}_{\dot\alpha A}^{CB} + \frac{1}{3} \overline{V}_{\dot\alpha\dot\gamma AD} \overline{W}^{\dot\gamma DCB} + \overline{S}_{AD} \overline{W}_{\dot\alpha}^{DCB} \right)$$

$$- i\, \mathcal{D}_{\dot\beta}^{\beta} \left(W_{\beta\alpha AD} \overline{W}_{\dot\alpha}^{DCB} \right) + 2\, U_{\alpha A}^{C\dot\gamma} \overline{W}_{\dot\gamma\dot\beta\dot\alpha}^{B} - \frac{1}{2} \delta_A^B U_{\alpha D}^{C\dot\gamma} \overline{W}_{\dot\gamma\dot\beta\dot\alpha}^{D}$$

$$- 4\, U_{\alpha\dot\beta D}^{C} \overline{\Lambda}_{\dot\alpha A}^{DB} + 2\,\overline{W}_{\dot\beta\dot\gamma}^{CB} \Psi^{\dot\gamma}{}_{\dot\alpha\alpha A} - 2\,\delta_A^C \overline{W}_{\dot\beta\dot\gamma}^{BD} \Psi^{\dot\gamma}{}_{\dot\alpha\alpha D}$$

$$+ 2\,\overline{W}_{\dot\beta\dot\alpha}^{CB} \Psi_{\alpha A} + 6\,\delta_A^C \overline{W}_{\dot\beta\dot\alpha}^{BD} \Psi_{\alpha D} - 2\,\delta_A^B \overline{W}_{\dot\beta\dot\alpha}^{CD} \Psi_{\alpha D}$$

$$+ 2\,\overline{W}_{\dot\beta\dot\alpha}^{CD} \Lambda_{\alpha DA}^{B} + \overline{M}_A^{CBD} \Psi_{\dot\beta\dot\alpha\alpha D} + \frac{2}{3} U_{\alpha\dot\beta D}^{C} \overline{V}_{\dot\alpha\dot\gamma AE} \overline{W}^{\dot\gamma EDB}$$

$$+ 2\, U_{\alpha\dot\beta D}^{C} \overline{S}_{AE} \overline{W}_{\dot\alpha}^{EDB} - V_{\alpha\beta}^{CD} \overline{W}_{\dot\beta\dot\alpha}^{BE} W_{EDA}^{\beta} - \frac{1}{3} \overline{W}_{\dot\beta\dot\alpha}^{CD} V_{\alpha\beta}^{BE} W_{EDA}^{\beta}$$

$$+ \overline{W}_{\dot\beta\dot\gamma}^{CD} \overline{W}^{\dot\gamma BE} W_{\alpha EDA} - \overline{W}_{\dot\beta\dot\alpha}^{CD} S^{BE} W_{\alpha EDA} + S^{CD} \overline{W}_{\dot\beta\dot\alpha}^{BE} W_{\alpha EDA}$$

$$+ W_{\alpha ADE} \overline{W}^{\dot\gamma DCB} \overline{W}_{\dot\gamma\dot\beta\dot\alpha}^{E} + 2\, W_{\alpha ADE} \overline{W}_{\dot\beta}^{DCB} \overline{\Psi}_{\dot\alpha}^{E} \bigg] \tag{E.6}$$

$$\sum_{\gamma\beta\alpha} \left(\mathcal{D}^{\delta}{}_{\dot\alpha} W_{\delta\gamma\beta\alpha} - \mathcal{D}_{\gamma}{}^{\dot\beta} P_{\beta\alpha\dot\beta\dot\alpha} + 3i\, \Psi_{\dot\alpha\dot\beta\gamma A} \overline{\Psi}_{\beta\alpha}^{A\,\dot\beta} \right.$$

$$\left. + 3i\, W_{\gamma\beta}{}^{\delta}{}_A \overline{\Psi}_{\delta\alpha\dot\alpha}^{A} + 9i\, W_{\gamma\beta\alpha A} \overline{\Psi}_{\dot\alpha}^{A} + 3i\, \Psi_{\gamma A} \overline{\Psi}_{\beta\alpha\dot\alpha}^{A} \right) = 0 \tag{E.7}$$

$$\mathcal{D}^{\beta\dot\beta} \Gamma_{\beta\alpha\dot\beta\dot\alpha} + \frac{1}{4} \mathcal{D}_{\alpha\dot\alpha} R + \frac{3}{2} i\, \left(W_{\alpha\beta\gamma A} \overline{\Psi}^{\beta\gamma A}{}_{\dot\alpha} + \overline{W}_{\dot\alpha\dot\beta\dot\gamma}^{A} \Psi_{\alpha A}^{\dot\beta\dot\gamma} \right)$$

$$- 6i\, \left(\Psi_A^{\beta} \overline{\Psi}_{\beta\alpha\dot\alpha}^{A} + \overline{\Psi}^{\dot\beta A} \Psi_{\dot\beta\dot\alpha\alpha A} \right) = 0 \tag{E.8}$$

$$\mathcal{D}^\beta_{\dot\alpha}\, \rho^B_{\beta\alpha A} + \mathcal{D}_\alpha{}^{\dot\beta}\, \bar\rho^B_{\dot\beta\dot\alpha A} - 2\mathrm{i}\left(W_{\alpha\beta\gamma A}\, \overline\Psi^{\beta\gamma\, B}{}_{\dot\alpha} + \overline W^B_{\dot\alpha\dot\beta\dot\gamma}\, \Psi^{\dot\beta\dot\gamma}_{\alpha A} \right)$$

$$+\frac{\mathrm{i}}{2}\,\delta^B_A\left(W_{\alpha\beta\gamma C}\, \overline\Psi^{\beta\gamma\, C}{}_{\dot\alpha} + \overline W^C_{\dot\alpha\dot\beta\dot\gamma}\, \Psi^{\dot\beta\dot\gamma}_{\alpha C} \right)$$

$$+2\mathrm{i}\left(\Psi_{\dot\alpha\dot\beta\,\alpha C}\, \overline\Lambda^{\dot\beta\, CB}_A + \overline\Psi^C_{\alpha\beta\dot\alpha}\, \Lambda^{\beta B}_{CA} \right)$$

$$+6\mathrm{i}\left(\Psi_{\alpha C}\, \overline\Lambda^{CB}_{\dot\alpha A} + \overline\Psi^C_{\dot\alpha}\, \Lambda^{B}_{\alpha CA} \right)$$

$$+\frac{\mathrm{i}}{3}\left(\Psi_{\dot\alpha\dot\beta\,\alpha C}\, \overline V^{\dot\beta\dot\gamma}_{AD}\, \overline W^{DCB}_{\dot\gamma} + \overline\Psi^C_{\alpha\beta\dot\alpha}\, V^{\beta\gamma\, BD}\, W_{\gamma DCA} \right)$$

$$-\mathrm{i}\left(\Psi_{\dot\alpha\dot\beta}{}^\beta{}_C\, W_{\beta\alpha AD}\, \overline W^{\dot\beta\, DCB} + \overline\Psi^C{}_{\alpha\beta}{}^{\dot\beta}\, \overline W^{BD}_{\dot\beta\dot\alpha}\, W^\beta_{DCA} \right)$$

$$-\mathrm{i}\left(\Psi_{\dot\alpha\dot\beta\,\alpha C}\, \overline S_{AD}\, \overline W^{\dot\beta\, DCB} + \overline\Psi^C_{\alpha\beta\dot\alpha}\, S^{BD}\, W^\beta_{DCA} \right)$$

$$-\mathrm{i}\left(W_{\alpha\beta\gamma C}\, W^{\beta\gamma}_{AD}\, \overline W^{DCB}_{\dot\alpha} + \overline W^C_{\dot\alpha\dot\beta\dot\gamma}\, \overline W^{\dot\beta\dot\gamma\, BD}\, W_{\alpha DCA} \right)$$

$$-\mathrm{i}\left(\Psi_{\alpha C}\, \overline V_{\dot\alpha\dot\beta AD}\, \overline W^{\dot\beta\, DCB} + \overline\Psi^C_{\dot\alpha}\, V^{BD}_{\alpha\beta}\, W^\beta_{DCA} \right)$$

$$+\mathrm{i}\left(\Psi^\beta_C\, W_{\beta\alpha AD}\, \overline W^{DCB}_{\dot\alpha} + \overline\Psi^{\dot\beta C}\, \overline W^{BD}_{\dot\beta\dot\alpha}\, W_{\alpha DCA} \right)$$

$$-3\mathrm{i}\left(\Psi_{\alpha C}\, \overline S_{AD}\, \overline W^{DCB}_{\dot\alpha} + \overline\Psi^C_{\dot\alpha}\, S^{BD}\, W_{\alpha DCA} \right) = 0 \qquad \text{(E.9)}$$

Appendix F

Non-Linear $N = 8$ Supergravity

Below we list the transformation laws, Bianchi identities, and field equations of $N = 8$ supergravity in superfield form. The linearized equations were already given in Section 12.1.

$$\mathcal{D}_{\dot\beta D} \overline{W}^{CBA}_{\dot\alpha} = \sum_{CBA} \delta^C_D \overline{W}^{BA}_{\dot\beta\dot\alpha} - \frac{1}{4} \varepsilon_{\dot\beta\dot\alpha} W^\alpha_{DEF} W^{CBAEF}_\alpha \tag{F.1}$$

$$\mathcal{D}^D_\alpha \overline{W}^{CBA}_{\dot\alpha} = i \overline{N}^{DCBA}_{\alpha\dot\alpha} \tag{F.2}$$

$$\mathcal{D}_{\dot\gamma C} \overline{W}^{BA}_{\dot\beta\dot\alpha} = -2 \sum_{BA} \delta^B_C \overline{W}^A_{\dot\gamma\dot\beta\dot\alpha} + \frac{1}{144} \sum_{\dot\gamma\dot\beta\dot\alpha} \overline{W}^{BAD}_{\dot\gamma} \overline{W}_{\dot\beta CDEFG} \overline{W}^{EFG}_{\dot\alpha}$$

$$+ \frac{1}{24} \sum_{\dot\beta\dot\alpha} \sum_{BA} \varepsilon_{\dot\gamma\dot\beta} \left(-\frac{3}{2} i W^\alpha_{CDE} \overline{N}^{BADE}_{\alpha\dot\alpha} - \frac{i}{3} \delta^B_C W^\alpha_{DEF} \overline{N}^{ADEF}_{\alpha\dot\alpha} \right.$$

$$\left. + \frac{1}{3} \overline{W}^{BAD}_{\dot\delta} \overline{W}^{\dot\delta}_{CDEFG} \overline{W}^{EFG}_{\dot\alpha} + \frac{1}{9} \delta^B_C \overline{W}^{ADE}_{\dot\delta} \overline{W}^{\dot\delta}_{DEFGH} \overline{W}^{FGH}_{\dot\alpha} \right) \tag{F.3}$$

$$\mathcal{D}^C_\alpha \overline{W}^{BA}_{\dot\beta\dot\alpha} = -\frac{i}{2} \sum_{\dot\beta\dot\alpha} \mathcal{D}_{\alpha\dot\beta} \overline{W}^{CBA}_{\dot\alpha}$$

$$- \frac{1}{8} \sum_{\dot\beta\dot\alpha} W_{\alpha DEF} \left(\overline{W}^{CDE}_{\dot\beta} \overline{W}^{BAF}_{\dot\alpha} + \frac{1}{6} \overline{W}^{CBA}_{\dot\beta} \overline{W}^{DEF}_{\dot\alpha} \right) \tag{F.4}$$

$$\mathcal{D}_{\dot\beta E} \overline{N}^{DCBA}_{\alpha\dot\alpha} = \frac{i}{4!} \sum_{\dot\beta\dot\alpha} \sum_{DCBA} \left(4i \delta^D_E \mathcal{D}_{\alpha\dot\beta} \overline{W}^{CBA}_{\dot\alpha} - W_{\alpha EFG} \overline{W}^{DCB}_{\dot\beta} \overline{W}^{AFG}_{\dot\alpha} \right.$$

$$\left. - \frac{3}{2} W_{\alpha EFG} \overline{W}^{DCF}_{\dot\beta} \overline{W}^{BAG}_{\dot\alpha} - \frac{1}{6} \delta^D_E W_{\alpha FGH} \overline{W}^{CBA}_{\dot\beta} \overline{W}^{FGH}_{\dot\alpha} \right)$$

$$+ \frac{i}{48} \varepsilon_{\dot\beta\dot\alpha} \sum_{DCBA} \left(W^\beta_{EFG} W^{DCBAFG}_{\beta\alpha} - 2 W^{\beta DCBAF} W_{\beta\alpha EF} \right.$$

$$- 4 W_{\alpha EFG} \overline{W}^{DCB}_{\dot\gamma} \overline{W}^{\dot\gamma AFG} - \delta^D_E W_{\alpha FGH} \overline{W}^{CBA}_{\dot\gamma} \overline{W}^{\dot\gamma FGH}$$

$$\left. + 3 \delta^D_E W_{\alpha FGH} \overline{W}^{CBF}_{\dot\gamma} \overline{W}^{\dot\gamma AGH} \right) \tag{F.5}$$

$$\overline{\Psi}^A_{\beta\alpha\dot\alpha} = \frac{1}{8} \sum_{\beta\alpha} \left(W_{\beta\alpha BC} \overline{W}^{ABC}_{\dot\alpha} - \frac{i}{6} W_{\beta BCD} \overline{N}^{ABCD}_{\alpha\dot\alpha} \right) \tag{F.6}$$

$$\overline{\Psi}{}^A_{\dot\alpha} = -\frac{1}{144}\left(\,\mathrm{i}\, W^\alpha_{BCD}\,\overline{N}{}^{ABCD}_{\alpha\dot\alpha} + \frac{1}{6}\,\overline{W}{}^{ABC}_{\dot\beta}\,\overline{W}{}^{\dot\beta}_{BCDEF}\,\overline{W}{}^{DEF}_{\dot\alpha}\right) \tag{F.7}$$

$$\mathcal{D}_{\alpha\dot\alpha}\overline{W}{}^{\dot\alpha CBA} = \frac{\mathrm{i}}{48}\sum_{CBA}\left(4\,W^{\beta CBADE}\,W_{\beta\alpha DE} + 3\,W_{\alpha DEF}\,\overline{W}{}^{CBD}_{\dot\alpha}\,\overline{W}{}^{\dot\alpha AEF}\right) \tag{F.8}$$

$$\mathcal{D}_{\dot\delta B}\,\overline{W}{}^A_{\dot\gamma\dot\beta\dot\alpha} = -\delta^A_B\,\overline{W}_{\dot\delta\dot\gamma\dot\beta\dot\alpha} + \frac{1}{72}\sum_{\dot\gamma\dot\beta\dot\alpha}\overline{W}{}^{AC}_{\dot\gamma\dot\beta}\,\overline{W}_{\dot\alpha BCDEF}\,\overline{W}{}^{DEF}_{\dot\delta}$$

$$+ \frac{1}{288}\sum_{\dot\gamma\dot\beta\dot\alpha}\varepsilon_{\dot\delta\dot\gamma}\Big[N^\alpha_{\dot\beta BCDE}\,\overline{N}{}^{ACDE}_{\alpha\dot\alpha} - 12\mathrm{i}\,W^\alpha_{BCD}\,\mathcal{D}_{\alpha\dot\beta}\overline{W}{}^{ACD}_{\dot\alpha}$$

$$+ \mathrm{i}\,\delta^A_B\,W^\alpha_{CDE}\,\mathcal{D}_{\alpha\dot\beta}\overline{W}{}^{CDE}_{\dot\alpha} + 3\,\overline{W}{}^{ACD}_{\dot\varepsilon}\,\overline{W}_{BCDEF}\,\overline{W}{}^{EF}_{\dot\beta\dot\alpha}$$

$$+ 2\,\overline{W}{}^{AC}_{\dot\beta\dot\varepsilon}\,\overline{W}{}^{\dot\varepsilon}_{BCDEF}\,\overline{W}{}^{DEF}_{\dot\alpha} - \frac{1}{2}\delta^A_B\,\overline{W}{}^{CD}_{\dot\beta\dot\varepsilon}\,\overline{W}{}^{\dot\varepsilon}_{CDEFG}\,\overline{W}{}^{EFG}_{\dot\alpha}$$

$$+ 3\,W^\alpha_{BCD}\,W_{\alpha EFG}\big(\overline{W}{}^{ACE}_{\dot\beta}\,\overline{W}{}^{DFG}_{\dot\alpha} - \frac{1}{2}\,\overline{W}{}^{AEF}_{\dot\beta}\,\overline{W}{}^{CDG}_{\dot\alpha}\big)\Big] \tag{F.9}$$

$$\mathcal{D}^B_\alpha\,\overline{W}{}^A_{\dot\gamma\dot\beta\dot\alpha} = \frac{1}{24}\sum_{\dot\gamma\dot\beta\dot\alpha}\Big[4\mathrm{i}\,\mathcal{D}_{\alpha\dot\gamma}\overline{W}{}^{BA}_{\dot\beta\dot\alpha} - \frac{\mathrm{i}}{6}\,N_{\alpha\dot\gamma CDEF}\,\overline{W}{}^{BAC}_{\dot\beta}\,\overline{W}{}^{DEF}_{\dot\alpha}$$

$$+ W_{\alpha CDE}\big(\overline{W}{}^{BAC}_{\dot\gamma}\,\overline{W}{}^{DE}_{\dot\beta\dot\alpha} - 2\,\overline{W}{}^{BCD}_{\dot\gamma}\,\overline{W}{}^{AE}_{\dot\beta\dot\alpha} - \frac{1}{3}\,\overline{W}{}^{CDE}_{\dot\gamma}\,\overline{W}{}^{BA}_{\dot\beta\dot\alpha}\big)\Big] \tag{F.10}$$

$$\sum_{\dot\beta\dot\alpha}\sum_{DCBA}\Big(\mathcal{D}^\alpha_{\dot\beta}\,\overline{N}{}^{DCBA}_{\alpha\dot\alpha} - 8\,\overline{W}{}^D_{\dot\beta\dot\alpha\dot\gamma}\,\overline{W}{}^{\dot\gamma CBA} - \overline{W}{}^{DE}_{\dot\beta\dot\alpha}\,W^\alpha_{EFG}\,W^{CBAFG}_\alpha$$

$$+ \frac{4}{9}\mathrm{i}\,\overline{W}{}^{DCB}_{\dot\beta}\,W^\alpha_{EFG}\,\overline{N}{}^{AEFG}_{\alpha\dot\alpha} + \frac{3}{2}\mathrm{i}\,\overline{W}{}^{DCE}_{\dot\beta}\,W^\alpha_{EFG}\,\overline{N}{}^{BAFG}_{\alpha\dot\alpha}$$

$$- \frac{5}{54}\,\overline{W}{}^{DCB}_{\dot\beta}\,\overline{W}{}^{AEF}_{\dot\gamma}\,\overline{W}{}^{\dot\gamma}_{EFGHK}\,\overline{W}{}^{GHK}_{\dot\alpha}$$

$$- \frac{1}{4}\,\overline{W}{}^{DCE}_{\dot\beta}\,\overline{W}{}^{BAF}_{\dot\gamma}\,\overline{W}{}^{\dot\gamma}_{EFGHK}\,\overline{W}{}^{GHK}_{\dot\alpha}\Big) = 0 \tag{F.11}$$

$$P_{\beta\alpha\dot\beta\dot\alpha} = \frac{1}{384}\sum_{\beta\alpha}\sum_{\dot\beta\dot\alpha}\Big[-48\,W_{\beta\alpha AB}\,\overline{W}{}^{AB}_{\dot\beta\dot\alpha} - \frac{1}{2}\,N_{\beta\dot\beta ABCD}\,\overline{N}{}^{ABCD}_{\alpha\dot\alpha}$$

$$+ 4\mathrm{i}\,W_{\beta ABC}\,\mathcal{D}_{\alpha\dot\alpha}\overline{W}{}^{ABC}_{\dot\beta} + 4\mathrm{i}\,\overline{W}{}^{ABC}_{\dot\beta}\,\mathcal{D}_{\alpha\dot\alpha}W_{\beta ABC}$$

$$+ W_{\beta ABC}\,W_{\alpha DEF}\Big(\frac{1}{6}\,\overline{W}{}^{ABC}_{\dot\beta}\,\overline{W}{}^{DEF}_{\dot\alpha} - 3\,\overline{W}{}^{ABD}_{\dot\beta}\,\overline{W}{}^{CEF}_{\dot\alpha}\Big)\Big] \tag{F.12}$$

$$R = \frac{1}{192} \left[N_{ABCD}^{\alpha\dot\alpha} \, \overline{N}_{\alpha\dot\alpha}^{ABCD} \right.$$

$$- 4 \, W_{AB}^{\alpha\beta} \, W_{\alpha}^{ABCDE} \, W_{\beta\,CDE} + 4 \, \overline{W}_{\dot\alpha\dot\beta}^{AB} \, \overline{W}_{ABCDE}^{\dot\alpha} \, \overline{W}^{\dot\beta\,CDE}$$

$$\left. + W_{ABC}^{\alpha} \, W_{\alpha\,DEF} \left(\overline{W}_{\dot\alpha}^{ABC} \, \overline{W}^{\dot\alpha\,DEF} - 6 \, \overline{W}_{\dot\alpha}^{ABD} \, \overline{W}^{\dot\alpha\,CEF} \right) \right] \tag{F.13}$$

$$\mathcal{D}_{\alpha}^{\ \dot\beta} \, \overline{W}_{\dot\beta\dot\alpha}^{BA} = \frac{\mathrm{i}}{96} \sum_{BA} \left[12\mathrm{i}\, W_{\alpha\ CD}^{\ \beta} \, \overline{N}_{\dot\beta\dot\alpha}^{BACD} + \mathrm{i}\, W^{\beta\,BACDE} \sum_{\alpha\beta} \mathcal{D}_{\alpha\dot\alpha} W_{\beta\,CDE} \right.$$

$$+ W_{\alpha\,CDE} \left(\overline{W}_{\dot\alpha\dot\beta}^{BA} \, \overline{W}^{\dot\beta\,CDE} + 6 \, \overline{W}_{\dot\alpha\dot\beta}^{BC} \, \overline{W}^{\dot\beta\,ADE} - 3 \, \overline{W}_{\dot\alpha\dot\beta}^{CD} \, \overline{W}^{\dot\beta\,BAE} \right)$$

$$- \mathrm{i}\, N_{\alpha\dot\alpha\,CDEF} \, \overline{W}_{\dot\beta}^{BAC} \, \overline{W}^{\dot\beta\,DEF} - 3\mathrm{i}\, N_{\alpha\dot\beta\,CDEF} \, \overline{W}^{\dot\beta\,BCD} \, \overline{W}_{\dot\alpha}^{AEF}$$

$$- \frac{11}{4} \, W_{\alpha}^{BACDE} \left(\frac{1}{3} \, W_{CDE}^{\beta} \, W_{\beta\,FGH} - \frac{3}{2} \, W_{CDF}^{\beta} \, W_{\beta\,EGH} \right) \overline{W}_{\dot\alpha}^{FGH}$$

$$\left. - W^{\beta\,BCDEF} \left(2 \, W_{\alpha\,CDE} \, W_{\beta\,FGH} - \frac{3}{2} \, W_{\alpha\,CDG} \, W_{\beta\,EFH} \right) \overline{W}_{\dot\alpha}^{AGH} \right] \tag{F.14}$$

$$\mathcal{D}^{\alpha\dot\alpha} \, \overline{N}_{\alpha\dot\alpha}^{DCBA} = \frac{1}{192} \sum_{DCBA} \left[8 \, W_{EF}^{\alpha\beta} \, W_{\alpha\beta}^{DCBAEF} + 96 \, \overline{W}_{\dot\alpha\dot\beta}^{DC} \, \overline{W}^{\dot\alpha\dot\beta\,BA} \right.$$

$$+ \frac{2}{3}\mathrm{i}\, W_{EFG}^{\alpha} \left(\overline{W}^{\dot\alpha\,EFG} \, \overline{N}_{\alpha\dot\alpha}^{DCBA} - 18 \, \overline{W}^{\dot\alpha\,DCE} \, \overline{N}_{\alpha\dot\alpha}^{BAFG} \right.$$

$$\left. - 8 \, \overline{W}^{\dot\alpha\,DCB} \, \overline{N}_{\alpha\dot\alpha}^{AEFG} \right) - \frac{1}{6} \, W_{EFG}^{\alpha} \, W_{HKL}^{\beta} \left(\frac{1}{6} \, W_{\alpha}^{DCBAH} \, W_{\beta}^{EFGKL} \right.$$

$$\left. - W_{\alpha}^{DCBHK} \, W_{\beta}^{AEFGL} + \frac{9}{4} \, W_{\alpha}^{DCEFH} \, W_{\beta}^{BAGKL} \right)$$

$$- \frac{9}{4} \, \overline{W}_{\dot\alpha}^{DCE} \, \overline{W}^{\dot\alpha\,BFG} \, \overline{W}_{\dot\beta}^{AHK} \, \overline{W}_{EFGHK}^{\dot\beta} - \frac{5}{2} \left(\overline{W}_{\dot\alpha}^{DCE} \, \overline{W}_{\dot\beta}^{BAF} \right.$$

$$\left. \left. + \frac{7}{18} \, \overline{W}_{\dot\alpha}^{DCB} \, \overline{W}_{\dot\beta}^{AEF} \right) \overline{W}^{\dot\alpha\,GHK} \, \overline{W}_{EFGHK}^{\dot\beta} \right] \tag{F.15}$$

$$\mathcal{D}_{\dot\epsilon A} \overline{W}_{\dot\delta\dot\gamma\dot\beta\dot\alpha} = \frac{1}{144} \sum_{\dot\delta\dot\gamma\dot\beta\dot\alpha} \left[\overline{W}_{\dot\epsilon\,ABCDE} \, \overline{W}_{\dot\delta}^{BCD} \, \overline{W}_{\dot\gamma\dot\beta\dot\alpha}^{E} + \frac{1}{2} \, \overline{W}_{\dot\delta\,ABCDE} \, \overline{W}_{\dot\gamma}^{BCD} \, \overline{W}_{\dot\epsilon\dot\beta\dot\alpha}^{E} \right.$$

$$- \varepsilon_{\dot\epsilon\dot\delta} \left(12\mathrm{i}\, \mathcal{D}^{\alpha}_{\ \dot\gamma} \, \Psi_{\dot\beta\dot\alpha\,\alpha A} + 3 \, W_{ABC}^{\alpha} \, \overline{W}_{\dot\gamma}^{BCD} \, \Psi_{\dot\beta\dot\alpha\,\alpha D} \right.$$

$$\left. \left. - \frac{1}{2} \, W_{BCD}^{\alpha} \, \overline{W}_{\dot\gamma}^{BCD} \, \Psi_{\dot\beta\dot\alpha\,\alpha A} + \overline{W}_{\dot\gamma\,ABCDE} \, \overline{W}_{\dot\beta}^{BCD} \, \overline{\Psi}_{\dot\alpha}^{E} \right) \right] \tag{F.16}$$

$$\mathcal{D}_{\alpha}^{A} \, \overline{W}_{\dot\delta\dot\gamma\dot\beta\dot\alpha} = \frac{1}{4!} \sum_{\dot\delta\dot\gamma\dot\beta\dot\alpha} \left(2\mathrm{i}\, \mathcal{D}_{\alpha\dot\delta} \, \overline{W}_{\dot\gamma\dot\beta\dot\alpha}^{A} + 4 \, \overline{W}_{\dot\delta\dot\gamma}^{AB} \, \Psi_{\dot\beta\dot\alpha\,\alpha B} \right.$$

$$\left. + \frac{3}{2} \, W_{\alpha\,BCD} \, \overline{W}_{\dot\delta}^{ABC} \, \overline{W}_{\dot\gamma\dot\beta\dot\alpha}^{D} - \frac{1}{4} \, W_{\alpha\,BCD} \, \overline{W}_{\dot\delta}^{BCD} \, \overline{W}_{\dot\gamma\dot\beta\dot\alpha}^{A} \right) \tag{F.17}$$

$$\mathcal{D}_\alpha{}^{\dot\gamma}\,\overline{W}{}^A_{\dot\gamma\dot\beta\dot\alpha} = \frac{1}{2}\sum_{\dot\beta\dot\alpha}\Big(\mathcal{D}^\beta{}_{\dot\beta}\,\overline{\Psi}{}^A_{\beta\alpha\dot\alpha} + \mathcal{D}_{\alpha\dot\beta}\,\overline{\Psi}{}^A_{\dot\alpha} + \mathrm{i}\,\overline{W}{}^{AB}_{\dot\beta\dot\gamma}\,\Psi{}^{\dot\gamma}_{\dot\alpha\,\alpha B} - 3\mathrm{i}\,\overline{W}{}^{AB}_{\dot\beta\dot\alpha}\,\Psi_{\alpha B}$$

$$+\,\frac{\mathrm{i}}{4}\,W_{\alpha BCD}\,\overline{W}{}^{\dot\gamma ABC}\,\overline{W}{}^D_{\dot\gamma\dot\beta\dot\alpha} - \frac{\mathrm{i}}{24}\,W_{\alpha BCD}\,\overline{W}{}^{\dot\gamma BCD}\,\overline{W}{}^A_{\dot\gamma\dot\beta\dot\alpha}$$

$$-\,\frac{\mathrm{i}}{24}\,W_\alpha^{ABCDE}\,W^\beta_{BCD}\,\Psi_{\dot\beta\dot\alpha\,\beta E} + \frac{\mathrm{i}}{2}\,W_{\alpha BCD}\,\overline{W}{}^{ABC}_{\dot\beta}\,\overline{\Psi}{}^D_{\dot\alpha}$$

$$-\,\frac{\mathrm{i}}{12}\,W_{\alpha BCD}\,\overline{W}{}^{BCD}_{\dot\beta}\,\overline{\Psi}{}^A_{\dot\alpha}\Big) \tag{F.18}$$

$$\mathcal{D}_\alpha{}^{\dot\delta}\,\overline{W}_{\dot\delta\dot\gamma\dot\beta\dot\alpha} = \frac{1}{3!}\sum_{\dot\gamma\dot\beta\dot\alpha}\Big(\mathcal{D}^\beta{}_{\dot\gamma}\,P_{\beta\alpha\dot\beta\dot\alpha} + 3\mathrm{i}\,\overline{W}{}^A_{\dot\gamma\dot\beta\dot\delta}\,\Psi{}^{\dot\delta}_{\dot\alpha\,\alpha A} - 9\mathrm{i}\,\overline{W}{}^A_{\dot\gamma\dot\beta\dot\alpha}\,\Psi_{\alpha A}$$

$$+\,3\mathrm{i}\,\Psi_{\dot\gamma\dot\beta\,A}{}^\beta\,\overline{\Psi}{}^A_{\beta\alpha\dot\alpha} + 3\mathrm{i}\,\Psi_{\dot\gamma\dot\beta\,\alpha A}\,\overline{\Psi}{}^A_{\dot\alpha}\Big) \tag{F.19}$$

References

1. J. Wess and J. Bagger, *Supersymmetry and supergravity* (Princeton Univ. Press, 1983)

2. S. J. Gates, Jr., M. T. Grisaru, M. Roček and W. Siegel, *Superspace* or *One thousand and one lessons in supersymmetry* (Benjamin/Cummings, 1983);
P. G. O. Freund, *Introduction to supersymmetry* (Cambridge Univ. Press, 1986);
P. P. Srivastava, *Supersymmetry, superfields and supergravity: an introduction* (Adam Hilger, Bristol, 1986);
P. West, *Introduction to supersymmetry and supergravity* (World Scientific, Singapore, 1986)

3. R. Haag, J. T. Łopuszański and M. Sohnius, *Nucl. Phys.* **B 88** (1975) 257

4. E. S. Fradkin and A. A. Tseytlin, *Phys. Rep.* **119** (1985) 233

5. M. Kaku, P. K. Townsend and P. van Nieuwenhuizen, *Phys. Lett.* **69 B** (1977) 304;
Phys. Rev. Lett. **39** (1977) 1109; *Phys. Rev.* **D 17** (1978) 3179;
P. K. Townsend and P. van Nieuwenhuizen, *Phys. Rev.* **D 19** (1979) 3166

6. B. de Wit and J. W. van Holten, *Nucl. Phys.* **B 155** (1979) 530;
B. de Wit, J. W. van Holten and A. Van Proeyen, *Nucl. Phys.* **B 167** (1980) 186 (E: **B 172** (1980) 543)

7. E. Bergshoeff, M. de Roo and B. de Wit, *Nucl. Phys.* **B 182** (1981) 173

8. B. de Wit and S. Ferrara, *Phys. Lett.* **81 B** (1979) 317

9. S. J. Gates, Jr. and R. Grimm, *Phys. Lett.* **133 B** (1983) 192

10. J. Wess and B. Zumino, *Phys. Lett.* **66 B** (1977) 361

11. P. Howe, *Phys. Lett.* **100 B** (1981) 389

12. P. Howe, *Nucl. Phys.* **B 199** (1982) 309

13. M. Müller, *Z. Phys.* **C 31** (1986) 321

14. N. Dragon, *Z. Phys.* **C 2** (1979) 29

15. P. S. Howe and R. W. Tucker, *Phys. Lett.* **80 B** (1978) 138

16. E. S. Fradkin and A. A. Tseytlin, *Nucl. Phys.* **B 203** (1982) 157

17. S.-C. Lee and P. van Nieuwenhuizen, *Phys. Rev.* **D 26** (1982) 934

18. W. Siegel, *Nucl. Phys.* **B 177** (1981) 325

19. B. Zumino, in *Recent developments in gravitation*, Cargèse 1978, eds. M. Lévy and S. Deser (Plenum, New York, 1979) p. 405

20. J. Wess and B. Zumino, *Phys. Lett.* **74 B** (1978) 51

21. E. Sokatchev, *Phys. Lett.* **100 B** (1981) 466

22. C. Ramirez, *Z. Phys.* **C 28** (1985) 281; **C 33** (1987) 455; *Ann. Phys.* **186** (1988) 43

23. J. Wess and B. Zumino, *Phys. Lett.* **79 B** (1978) 394

24. P. van Nieuwenhuizen, *Phys. Rep.* **68** (1981) 189

25. D. Z. Freedman, P. van Nieuwenhuizen and S. Ferrara, *Phys. Rev.* **D 13** (1976) 3214;
S. Deser and B. Zumino, *Phys. Lett.* **62 B** (1976) 335

26. S. Ferrara and P. van Nieuwenhuizen, *Phys. Rev. Lett.* **37** (1976) 1669

27. D. Z. Freedman, *Phys. Rev. Lett.* **38** (1977) 105;
    S. Ferrara, J. Scherk and B. Zumino, *Phys. Lett.* **66 B** (1977) 35

28. A. Das, *Phys. Rev.* **D 15** (1977) 2805;
    E. Cremmer and J. Scherk, *Nucl. Phys.* **B 127** (1977) 259

29. E. Cremmer, J. Scherk and S. Ferrara, *Phys. Lett.* **74 B** (1978) 61;
    H. Nicolai and P. K. Townsend, *Phys. Lett.* **98 B** (1981) 257

30. B. de Wit and D. Z. Freedman, *Nucl. Phys.* **B 130** (1977) 105

31. E. Cremmer and B. Julia, *Phys. Lett.* **80 B** (1978) 48; *Nucl. Phys.* **B 159** (1979) 141

32. C. Aragone and S. Deser, *Phys. Lett.* **86 B** (1979) 161;
    F. A. Berends, J. W. van Holten, B. de Wit and P. van Nieuwenhuizen, *J. Phys.* **A 13** (1980) 1643;
    T. Curtright, *Phys. Lett.* **85 B** (1979) 219

33. P. Breitenlohner, *Phys. Lett.* **67 B** (1977) 49; *Nucl. Phys.* **B 124** (1977) 500;
    W. Siegel and S. J. Gates, Jr., *Nucl. Phys.* **B 147** (1979) 77

34. S. Ferrara and P. van Nieuwenhuizen, *Phys. Lett.* **74 B** (1978) 333;
    K. S. Stelle and P. C. West, *Phys. Lett.* **74 B** (1978) 330

35. B. de Wit and P. van Nieuwenhuizen, *Nucl. Phys.* **B 139** (1978) 216

36. M. F. Sohnius and P. C. West, *Phys. Lett.* **105 B** (1981) 353

37. V. O. Rivelles and J. G. Taylor, *Phys. Lett.* **113 B** (1982) 467

38. G. Girardi, R. Grimm, M. Müller and J. Wess, *Phys. Lett.* **147 B** (1984) 81

39. E. S. Fradkin and M. A. Vasiliev, *Lett. Nuovo Cim.* **25** (1979) 79; *Phys. Lett.* **85 B** (1979) 47

40. B. de Wit, J. W. van Holten and A. Van Proeyen, *Nucl. Phys.* **B 184** (1981) 77

41. B. de Wit, R. Philippe and A. Van Proeyen, *Nucl. Phys.* **B 219** (1983) 143

42. M. Müller, *Phys. Lett.* **B 172** (1986) 353

43. M. Müller, *Nucl. Phys.* **B 282** (1987) 329; **B 289** (1987) 557

44. A. Galperin, E. Ivanov, V. Ogievetsky and E. Sokatchev, *Class. Quantum Grav.* **4** (1987) 1255

45. V. O. Rivelles and J. G. Taylor, *Phys. Lett.* **104 B** (1981) 131; **121 B** (1983) 37;
    J. G. Taylor, *J. Phys.* **A 15** (1982) 867

46. M. F. Sohnius, *Nucl. Phys.* **B 138** (1978) 109

47. A. Galperin, E. Ivanov, S. Kalitzin, V. Ogievetsky and E. Sokatchev, *Class. Quantum Grav.* **1** (1984) 469

48. W. Siegel, *Class. Quantum Grav.* **3** (1986) L 47;
    M. Hayashi and S. Uehara, *Phys. Lett.* **B 172** (1986) 348;
    C. S. Aulakh, J.-P. Derendinger and S. Ouvry, *Phys. Lett.* **169 B** (1986) 201

49. K. S. Stelle and P. C. West, *Nucl. Phys.* **B 145** (1978) 175

50. T. Kugo and S. Uehara, *Nucl. Phys.* **B 226** (1983) 49, 93

51. S. Ferrara and C. A. Savoy, in *Supergravity '81*, Trieste 1981, eds. S. Ferrara and J. G. Taylor (Cambridge Univ. Press, 1982) p. 47

52. M. F. Sohnius, *Phys. Rep.* **128** (1985) 39

53. L. Brink and P. Howe, *Phys. Lett.* **88 B** (1979) 268

54. M. de Roo and P. Wagemans, *Nucl. Phys.* **B 262** (1985) 644

55. V. I. Ogievetsky and I. V. Polubarinov, *Sov. J. Nucl. Phys.* **4** (1967) 156

56. P. Breitenlohner and M. F. Sohnius, *Nucl. Phys.* **B 165** (1980) 483; **B 178** (1981) 151

57. M. Müller, *Z. Phys.* **C 24** (1984) 175

58. P. S. Howe, K. S. Stelle and P. K. Townsend, *Phys. Lett.* **107 B** (1981) 420;
    S. J. Gates, Jr., M. Roček and W. Siegel, *Nucl. Phys.* **B 198** (1982) 113

59. E. Cremmer, S. Ferrara, L. Girardello and A. Van Proeyen, *Phys. Lett.* **116 B** (1982) 231; *Nucl. Phys.* **B 212** (1983) 413

60. W. Lang, J. Louis and B. A. Ovrut, *Nucl. Phys.* **B 261** (1985) 334;
    R. Garreis and C. Schwiebert, *Nucl. Phys.* **B 296** (1988) 902

61. J. Wess and B. Zumino, *Nucl. Phys.* **B 70** (1974) 39

62. W. Siegel, *Phys. Lett.* **85 B** (1979) 333

63. S. J. Gates, Jr., *Nucl. Phys.* **B 184** (1981) 381

64. B. de Wit and M. Roček, *Phys. Lett.* **109 B** (1982) 439

65. E. Cremmer, B. Julia, J. Scherk, P. van Nieuwenhuizen, S. Ferrara and L. Girardello, *Phys. Lett.* **79 B** (1978) 231; *Nucl. Phys.* **B 147** (1979) 105

66. E. Witten and J. Bagger, *Phys. Lett.* **115 B** (1982) 202

67. J. A. Bagger, *Nucl. Phys.* **B 211** (1983) 302

68. T. Kugo and S. Uehara, *Nucl. Phys.* **B 222** (1983) 125

69. B. Zumino, *Phys. Lett.* **87 B** (1979) 203

70. P. Binétruy, G. Girardi, R. Grimm and M. Müller, *Phys. Lett.* **B 189** (1987) 83

71. D. Z. Freedman, *Phys. Rev.* **D 15** (1977) 1173

72. K. S. Stelle and P. C. West, *Nucl. Phys.* **B 145** (1978) 175

73. M. Sohnius and P. West, *Nucl. Phys.* **B 203** (1982) 179;
    R. Barbieri, S. Ferrara, D. V. Nanopoulos and K. S. Stelle, *Phys. Lett.* **113 B** (1982) 219

74. M. Müller, *Nucl. Phys.* **B 264** (1986) 292

75. S. Samuel, *Nucl. Phys.* **B 245** (1984) 127

76. S. Ferrara, L. Girardello, T. Kugo and A. Van Proeyen, *Nucl. Phys.* **B 223** (1983) 191

77. P. Fayet, *Nucl. Phys.* **B 113** (1976) 135

78. P. S. Howe, K. S. Stelle and P. K. Townsend, *Nucl. Phys.* **B 214** (1983) 519

79. J. P. Yamron and W. Siegel, *Nucl. Phys.* **B 263** (1986) 70

80. R. Grimm, M. Sohnius and J. Wess, *Nucl. Phys.* **B 133** (1978) 275

81. M. Sohnius, K. S. Stelle and P. C. West, *Phys. Lett.* **92 B** (1980) 123; *Nucl. Phys.* **B 173** (1980) 127

82. U. Lindström and M. Roček, *Nucl. Phys.* **B 222** (1983) 285

83. J. Bagger and E. Witten, *Nucl. Phys.* **B 222** (1983) 1

84. B. de Wit, P. G. Lauwers, R. Philippe, Su, S.-Q. and A. Van Proeyen, *Phys. Lett.* **134 B** (1984) 37

85. B. de Wit, P. G. Lauwers and A. Van Proeyen, *Nucl. Phys.* **B 255** (1985) 569

86. K. Galicki, *Nucl. Phys.* **B 271** (1986) 402; **B 289** (1987) 573

87. J. A. Bagger, A. S. Galperin, E. A. Ivanov and V. I. Ogievetsky, *Nucl. Phys.* **B 303** (1988) 522

88. B. de Wit and A. Van Proeyen, *Nucl. Phys.* **B 245** (1984) 89

89. A. Galperin, E. Ivanov, S. Kalitzin, V. Ogievetsky and E. Sokatchev, *Class. Quantum Grav.* **2** (1985) 155

90. L. Castellani, A. Ceresole, R. D'Auria, S. Ferrara, P. Fré and E. Maina, *Phys. Lett.* **161 B** (1985) 91; *Nucl. Phys.* **B 268** (1986) 317

91. D. Z. Freedman and A. Das, *Nucl. Phys.* **B 120** (1977) 221

92. A. Das, M. Fischler and M. Roček, *Phys. Rev.* **D 16** (1977) 3427;
    D. Z. Freedman and J. H. Schwarz, *Nucl. Phys.* **B 137** (1978) 333;
    S. J. Gates, Jr. and B. Zwiebach, *Phys. Lett.* **123 B** (1983) 200

93. B. de Wit and H. Nicolai, *Nucl. Phys.* **B 188** (1981) 98

94. B. de Wit and H. Nicolai, *Phys. Lett.* **108 B** (1982) 285; *Nucl. Phys.* **B 208** (1982) 323

95. P. Howe and H. Nicolai, *Phys. Lett.* **109 B** (1982) 269

96. C. M. Hull, *Class. Quantum Grav.* **2** (1985) 343

97. G. 't Hooft and M. Veltman, *Ann. Inst. H. Poincaré* **20** (1974) 69

98. M. H. Goroff and A. Sagnotti, *Phys. Lett.* **160 B** (1985) 81; *Nucl. Phys.* **B 266** (1986) 709

99. M. T. Grisaru, *Phys. Lett.* **66 B** (1977) 75;
    E. Tomboulis, *Phys. Lett.* **67 B** (1977) 417

100. P. Howe and U. Lindström, *Nucl. Phys.* **B 181** (1981) 487;
     R. E. Kallosh, *Phys. Lett.* **99 B** (1981) 122;
     P. S. Howe, K. S. Stelle and P. K. Townsend, *Nucl. Phys.* **B 191** (1981) 445

101. K. S. Stelle, *Phys. Rev.* **D 16** (1977) 953

102. K. S. Stelle, *Gen. Rel. Grav.* **9** (1978) 353

103. S. Ferrara, M. T. Grisaru and P. van Nieuwenhuizen, *Nucl. Phys.* **B 138** (1978) 430

104. R. Grimm, Ph. D. thesis, Univ. Karlsruhe, 1979

105. B. de Wit, *Nucl. Phys.* **B 158** (1979) 189

Super-Index

This index contains the most frequently occuring superfields. The brackets indicate their symmetry properties (see Appendix A) and the numbers refer to the pages where they are defined.

$B^{[BA]}$	71	$\overline{\Lambda}^{[BA]}_{\dot{\alpha}C}$	13	$\overline{\Psi}^{A}_{\dot{\alpha}}$	13
$\overline{B}^{A}_{\dot{\alpha}}$	71	$\overline{\Lambda}_{\dot{\alpha}A}$	73	R	14
E	5	$\overline{M}^{[CBA]}_{D}$	11	$\overline{\rho}^{B}_{(\dot{\beta}\dot{\alpha})A}$	14
$\mathcal{E}$	43	$\overline{M}^{[DCBA]}_{\alpha E}$	25, 108	$S^{(BA)}$	11
$F^{[BA]}$	57–58	$\overline{M}^{[EDCBA]}_{(\beta\alpha)F}$	27, 111	$U^{B}_{\alpha\dot{\alpha}A}$	12
F	73	$\overline{N}^{[DCBA]}_{\alpha\dot{\alpha}}$	12	$U_{\alpha\dot{\alpha}}$	74
$F^{[CBA]}_{\alpha}$	58	$\overline{N}^{[EDCBA]}_{(\beta\alpha)\dot{\alpha}}$	25, 108	$V^{[BA]}_{(\beta\alpha)}$	11
$\overline{F}^{A}_{\dot{\alpha}}$	58	$P_{(\beta\alpha)(\dot{\beta}\dot{\alpha})}$	14	$V_{(\beta\alpha)}$	73
$\overline{F}_{(\dot{\beta}\dot{\alpha})}$	58	$P^{[BA]}_{[DC]}$	27, 109	$\overline{W}^{[DCBA]}$	31, 52
F^{A}_{B}	59	P	73	W_{i}	34
G^{A}_{B}	69	Φ	37, 76	$\overline{W}^{[CBA]}_{\dot{\alpha}}$	10
G^{A}_{α}	69	φ	77, 79	$\overline{W}^{[BA]}_{(\dot{\beta}\dot{\alpha})}$	11
$G^{[BA]}$	69	φ^{A}_{α}	75–76, 79	$\overline{W}_{(\dot{\beta}\dot{\alpha})}$	73
G	79	$\varphi_{\alpha\dot{\alpha}}$	76	$\overline{W}^{A}_{(\dot{\gamma}\dot{\beta}\dot{\alpha})}$	13
$\widetilde{G}_{\alpha\dot{\alpha}}$	69	$\varphi^{(BA)}_{\alpha\dot{\alpha}}$	79	$\overline{W}_{(\dot{\delta}\dot{\gamma}\dot{\beta}\dot{\alpha})}$	14
H	20–21	$\overline{\Psi}^{A}_{(\beta\alpha)\dot{\alpha}}$	13		

This article was processed by the author using the TeX Macropackage from Springer-Verlag.

Lecture Notes in Mathematics

Vol. 1236: Stochastic Partial Differential Equations and Applications. Proceedings, 1985. Edited by G. Da Prato and L. Tubaro. V, 257 pages. 1987.

Vol. 1237: Rational Approximation and its Applications in Mathematics and Physics. Proceedings, 1985. Edited by J. Gilewicz, M. Pindor and W. Siemaszko. XII, 350 pages. 1987.

Vol. 1250: Stochastic Processes – Mathematics and Physics II. Proceedings 1985. Edited by S. Albeverio, Ph. Blanchard and L. Streit. VI, 359 pages. 1987.

Vol. 1251: Differential Geometric Methods in Mathematical Physics. Proceedings, 1985. Edited by P.L. García and A. Pérez-Rendón. VII, 300 pages. 1987.

Vol. 1255: Differential Geometry and Differential Equations. Proceedings, 1985. Edited by C. Gu, M. Berger and R.L. Bryant. XII, 243 pages. 1987.

Vol. 1256: Pseudo-Differential Operators. Proceedings, 1986. Edited by H.O. Cordes, B. Gramsch and H. Widom. X, 479 pages. 1987.

Vol. 1258: J. Weidmann, Spectral Theory of Ordinary Differential Operators. VI, 303 pages. 1987.

Vol. 1260: N.H. Pavel, Nonlinear Evolution Operators and Semigroups. VI, 285 pages. 1987.

Vol. 1263: V.L. Hansen (Ed.), Differential Geometry. Proceedings, 1985. XI, 288 pages. 1987.

Vol. 1265: W. Van Assche, Asymptotics for Orthogonal Polynomials. VI, 201 pages. 1987.

Vol. 1267: J. Lindenstrauss, V.D. Milman (Eds.), Geometrical Aspects of Functional Analysis. Seminar. VII, 212 pages. 1987.

Vol. 1269: M. Shiota, Nash Manifolds. VI, 223 pages. 1987.

Vol. 1270: C. Carasso, P.-A. Raviart, D. Serre (Eds.), Nonlinear Hyperbolic Problems. Proceedings, 1986. XV, 341 pages. 1987.

Vol. 1272: M.S. Livšic, L.L. Waksman, Commuting Nonselfadjoint Operators in Hilbert Space. III, 115 pages. 1987.

Vol. 1273: G.-M. Greuel, G. Trautmann (Eds.), Singularities, Representation of Algebras, and Vector Bundles. Proceedings, 1985. XIV, 383 pages. 1987.

Vol. 1275: C.A. Berenstein (Ed.), Complex Analysis I. Proceedings, 1985–86. XV, 331 pages. 1987.

Vol. 1276: C.A. Berenstein (Ed.), Complex Analysis II. Proceedings, 1985–86. IX, 320 pages. 1987.

Vol. 1277: C.A. Berenstein (Ed.), Complex Analysis III. Proceedings, 1985–86. X, 350 pages. 1987.

Vol. 1283: S. Mardešić, J. Segal (Eds.), Geometric Topology and Shape Theory. Proceedings, 1986. V, 261 pages. 1987.

Vol. 1285: I.W. Knowles, Y. Saitō (Eds.), Differential Equations and Mathematical Physics. Proceedings, 1986. XVI, 499 pages. 1987.

Vol. 1287: E.B. Saff (Ed.), Approximation Theory, Tampa. Proceedings, 1985–1986. V, 228 pages. 1987.

Vol. 1288: Yu. L. Rodin, Generalized Analytic Functions on Riemann Surfaces. V, 128 pages, 1987.

Vol. 1294: M. Queffélec, Substitution Dynamical Systems – Spectral Analysis. XIII, 240 pages. 1987.

Vol. 1299: S. Watanabe, Yu.V. Prokhorov (Eds.), Probability Theory and Mathematical Statistics. Proceedings, 1986. VIII, 589 pages. 1988.

Vol. 1300: G.B. Seligman, Constructions of Lie Algebras and their Modules. VI, 190 pages. 1988.

Vol. 1302: M. Cwikel, J. Peetre, Y. Sagher, H. Wallin (Eds.), Function Spaces and Applications. Proceedings, 1986. VI, 445 pages. 1988.

Vol. 1303: L. Accardi, W. von Waldenfels (Eds.), Quantum Probability and Applications III. Proceedings, 1987. VI, 373 pages. 1988.

Lecture Notes in Physics

Vol. 312: J.S. Feldman, Th.R. Hurd, L. Rosen, "QED: A Proof of Renormalizability." VII, 176 pages. 1988.

Vol. 313: H.-D. Doebner, T.D. Palev, J.D. Hennig (Eds.), Group Theoretical Methods in Physics. Proceedings, 1987. XI, 599 pages. 1988.

Vol. 314: L. Peliti, A. Vulpiani (Eds.), Measures of Complexity. Proceedings, 1987. VII, 150 pages. 1988.

Vol. 315: R.L. Dickman, R.L. Snell, J.S. Young (Eds.), Molecular Clouds in the Milky Way and External Galaxies. Proceedings, 1987. XVI, 475 pages. 1988.

Vol. 316: W. Kundt (Ed.), Supernova Shells and Their Birth Events. Proceedings, 1988. VIII, 253 pages. 1988.

Vol. 317: C. Signorini, S. Skorka, P. Spolaore, A. Vitturi (Eds.), Heavy Ion Interactions Around the Coulomb Barrier. Proceedings, 1988. X, 329 pages. 1988.

Vol. 318: B. Mercier, An Introduction to the Numerical Analysis of Spectral Methods. V, 154 pages. 1989.

Vol. 319: L. Garrido (Ed.), Far from Equilibrium Phase Transitions. Proceedings, 1988. VIII, 340 pages. 1988.

Vol. 320: D. Coles (Ed.), Perspectives in Fluid Mechanics. Proceedings, 1985. VII, 207 pages. 1988.

Vol. 321: J. Pitowsky, Quantum Probability – Quantum Logic. IX, 209 pages. 1989.

Vol. 322: M. Schlichenmaier, An Introduction to Riemann Surfaces, Algebraic Curves and Moduli Spaces. XIII, 148 pages. 1989.

Vol. 323: D.L. Dwoyer, M.Y. Hussaini, R.G. Voigt (Eds.), 11th International Conference on Numerical Methods in Fluid Dynamics. XIII, 622 pages. 1989.

Vol. 324: P. Exner, P. Šeba (Eds.), Applications of Self-Adjoint Extensions in Quantum Physics. Proceedings, 1987. VIII, 273 pages. 1989.

Vol. 325: E. Brändas, N. Elander (Eds.), Resonances, Proceedings, 1987. XVIII, 564 pages. 1989.

Vol. 327: K. Meisenheimer, H.-J. Röser (Eds.), Hot Spots in Extragalactic Radio Source. Proceedings, 1988. XII, 301 pages. 1989.

Vol. 328: G. Wegner (Ed.), White Dwarfs. Proceedings, 1988. XIV, 524 pages. 1989.

Vol. 329: A. Heck, F. Murtagh (Eds.), Knowledge Based Systems in Astronomy. IV, 280 pages. 1989.

Vol. 330: J.M. Moran, J.N. Hewitt, K.Y. Lo (Eds.), Gravitational Lenses. Proceedings, 1988. XIV, 238 pages. 1989.

Vol. 331: G. Winnewisser, J.T. Armstrong (Eds.), The Physics and Chemistry of Interstellar Molecular Clouds mm and Sub-mm Observations in Astrophysics. Proceedings, 1988. XVIII, 463 pages. 1989.

Vol. 332: P. Flin, H.W. Duerbeck (Eds.), Morphological Cosmology. Proceedings, 1988. VII, 438 pages. 1989.

Vol. 333: I. Appenzeller, H.J. Habing, P. Léna (Eds.), Evolution of Galaxies – Astronomical Observations. Proceedings, 1988. X, 391 pages. 1989.

Vol. 335: A. Lakhtakia, V.K. Varadan, V.V. Varadan, Time-Harmonic Electromagnetic Fields in Chiral Media. VII, 121 pages. 1989.

Vol. 336: M. Müller, Consistent Classical Supergravity Theories. VI, 125 pages. 1989.